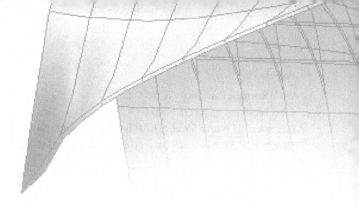

A Pragmatic Introduction to

the Finite Element Method for

Thermal and Stress Analysis

With the Matlab Toolkit SOFEA

Petr Krysl

University of California, San Diego

T0331571

 **World Scientific**

NEW JERSEY · LONDON · SINGAPORE · BEIJING · SHANGHAI · HONG KONG · TAIPEI · CHENNAI

Published by

World Scientific Publishing Co. Pte. Ltd.

5 Toh Tuck Link, Singapore 596224

USA office: 27 Warren Street, Suite 401-402, Hackensack, NJ 07601

UK office: 57 Shelton Street, Covent Garden, London WC2H 9HE

British Library Cataloguing-in-Publication Data
A catalogue record for this book is available from the British Library.

A PRAGMATIC INTRODUCTION TO THE FINITE ELEMENT METHOD
FOR THERMAL AND STRESS ANALYSIS
With the Matlab Toolkit SOFEA

ISBN-13 978-981-256-876-2
ISBN-10 981-256-876-X
ISBN-13 978-981-270-411-5 (pbk)
ISBN-10 981-270-411-6 (pbk)

Printed in Singapore

Preface

This book focuses on the two continuum mechanics models that are commonly encountered by mechanical, aerospace, civil, chemical, bio, and manufacturing engineering students: heat conduction in solids, and stress analysis. Its main purpose is to provide the reader with an insight into the workings of the continuum models and the finite element method by supplying information sufficient to guide intelligent modeling while avoiding tedious formal theorem and proof sequences. The initial boundary value problems are presented with more care than usually found in textbooks at this level, and in particular a proper treatment of the boundary conditions is given much attention. The basis for the formulation of the finite element models is the Galerkin method, as a special case of the method of weighted residuals. This is a very general approach, more broadly applicable than techniques based on variational principles, and it was chosen with the hope of serving the students well throughout their academic careers, including graduate-level courses on numerical solutions of nonlinear initial boundary value problems.

This book is a precipitate of lectures given over the years to Structural Engineering majors in their senior year at the University of California, San Diego. There are two aspects to the book: the first and foremost is a gradual and rational construction of the framework of the finite element models; the second, for the most part parallel, but sometimes subordinate, is the programming of the discussed algorithms in a sound software-engineering methodology. The first aspect of the book is comfortably covered in an undergraduate course in one quarter, but the implementation is only touched upon here and there. One semester would allow ample time for full indepth treatment of both aspects. On the other hand, presenting this book to graduate students who have been exposed to finite elements before would allow for the entire book to be studied thoroughly in one quarter, with equal coverage of both aspects.

The students should have a working knowledge of multivariable calculus, differential equations, and linear algebra. The more advanced mathematical tools are reviewed when and where needed. Familiarity with the the basics of solid mechanics will be helpful, but since no important steps are being skipped in the formulations of the models, the book is really practically self-contained in this respect.

An important characteristic of the book is its pragmatic slant: Not only being comprehensible was more important to me than mathematical rigor, but I endeavored to link the book to a software framework that would allow for hands-on, DIY experimenting at all levels of the discussed methods and algorithms. This important resource is the object-oriented Matlab toolbox SOFEA, freely available from the author, including all updates and corrections, at

```
http://hogwarts.ucsd.edu/~pkrysl/sofea
```

The toolbox implements all the models discussed in this book, and is quite useful for research experiments too. Some of the extensions of this toolbox to other problems are also available on the above web site. SOFEA does not require any other software other than Matlab itself. It has been tested with Release 14, but even earlier releases are able to run the majority of SOFEA classes and example scripts.

I would like to extend my warmest appreciation to those numerous readers of the various drafts of this book, including my students and the anonymous reviewers, who pointed out typos, omissions, and suggested numerous improvements. Thanks!

I hope you have at least as much fun reading the book as I had writing it.

Petr Krysl

Contents

Chapter 1

Model of a Taut Wire

This chapter will formulate a relatively simple model (a so-called initial boundary value problem) that describes the deflection or vibrations of a taut string. In the next chapter, we will seek approximate solutions to this model with the Galerkin method.

1.1 Deriving the PDE model

Figure 1.1 illustrates an idealization of a taut wire. The wire is under prestress by the force P, assumed to be uniform along the length of the wire. The left-hand end is immovably fixed, while the right-hand end is held in a fixture which can slide perpendicularly to the axis of the wire. A transverse force F_L is applied at the movable end. In addition, there may be some distributed force q acting along the length (for instance gravity). The transverse displacement is a function of both the axial coordinate x and the time t, $w = w(x, t)$. The transverse displacement is assumed to be very small compared to the length of the wire.

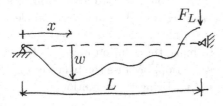

Fig. 1.1   Schematic of taut wire.

1

1.2   Balance equation

Taking a section of length Δx of the wire (see Fig. 1.2), collecting all the forces, and equating them to the inertial force (Newton's law), leads to a **balance equation** for the taut wire

$$P\frac{\partial^2 w}{\partial x^2} + q = \mu\ddot{w} \ , \tag{1.1}$$

where $\ddot{w} = \dfrac{\partial^2 w}{\partial t^2}$ is the acceleration.

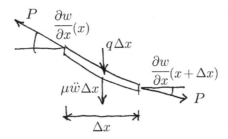

Fig. 1.2   The forces acting on a segment of the taut wire.

1.3   Boundary conditions

The function w that describes the transverse deflection takes two arguments, x, and t. It is defined on a rectangle shown in Fig. 1.3: $0 \leq x \leq L$, and $0 \leq t \leq \bar{t}$. It needs to be determined to satisfy the balance equation (1.1), but that would not completely nail the answer down. Indeed, there are other things we would require a solution to satisfy, namely the conditions at the **boundaries** of the domain rectangle.

How many pieces of information do we need to know? A reasonable answer is, 'Enough to make the solution unique.' To find the deflection w is going to involve integration, because the balance equation refers to space and time derivatives of w. Using the definitions

$$v = \frac{\partial w}{\partial t} \ , \qquad \theta = \frac{\partial w}{\partial x} \ ,$$

we may rewrite the balance equation that involves the second derivatives of the function w as a system of first order partial differential equations

$$\frac{\partial \theta}{\partial t} = \frac{\partial v}{\partial x}$$

$$P\frac{\partial \theta}{\partial x} + q - \mu\frac{\partial v}{\partial t} = 0$$

For each derivative $\frac{\partial v}{\partial x}$, $\frac{\partial \theta}{\partial x}$, one boundary condition (integration constant) will be needed. Similarly, for each of the time derivatives $\frac{\partial v}{\partial t}$, and $\frac{\partial \theta}{\partial t}$ one boundary condition along the time axis will be required.

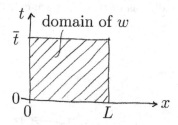

Fig. 1.3   The domain of the deflection function w.

1.4   Boundary conditions (in space)

The conditions on w along the edges of the domain rectangle parallel to the time axis are known (for historical reasons) as *the **boundary conditions***. (Perhaps also because they are applied along the *physical boundaries* of the structure.)

It needs to be realized that the domain of the wire, that is the interval $0 \le x \le L$, has only one boundary, namely the two endpoints, $x = 0$ and $x = L$. Since these two points are disjoint, the boundary of the interval consists of two disjoint sets. As discussed in more detail in Chapter 4, we are really prescribing a single boundary condition. Since it happens to be applied at two disjoint points, we loosely use the plural "boundary conditions".

In this example, at the left-hand end of the wire we are prescribing in general nonzero displacement,

$$w(0, t) = \bar{w}_0(t) \, . \tag{1.2}$$

As we shall find out, there is a good reason why this kind of condition is commonly called the **essential boundary condition.**

At the other end the boundary condition is of a different nature. It is also a bit more interesting, as we have to derive it. Again, we take a short section of the wire of length Δx (see Fig. 1.4). This time there are terms that are multiplied by Δx, but there are also others which are not. Only the latter survive when we make Δx go to zero

$$-P\frac{\partial w}{\partial x}(L,t) + F_L(t) = 0 \ . \tag{1.3}$$

This boundary condition is simply the balance of forces at the end of the wire. Boundary conditions of this kind are called **natural boundary conditions.**

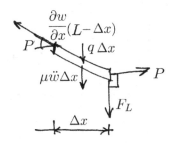

Fig. 1.4   The forces acting on the right-hand end of the taut wire.

1.5   Initial conditions (boundary conditions in time)

Along the edges of the domain rectangle that are parallel to the space axis we also apply two pieces of information. However, as we are all aware, the time direction is special. Therefore, it will probably come to us naturally to expect to know something about the deflection at *one point* in time, typically at $t = 0$. Because this is the initial point along the time axis, this condition is known as the **initial condition** (and we need two equations, one for each variable or for each derivative, in order to compensate for prescribing the condition at one point only):

$$w(x,0) = \bar{W}(x), \quad \frac{\partial w}{\partial t}(x,0) = \bar{V}(x) \ , \tag{1.4}$$

where $\bar{W}(x)$ (the initial deflection) and $\bar{V}(x)$ (the initial velocity) are known functions.

1.6 Anything else?

The balance equation (1.1), the boundary conditions (1.2) and (1.3), and the initial conditions (1.4) are all we need to fully define what model it is we are trying to find solutions to. It is an *initial boundary value problem*, and as such it is quite typical of the models with which structural engineers have to deal. In what follows, we shall find out how to formulate an algorithm, the so-called Galerkin finite element method, which will supply an approximate solution to this problem.

Chapter 2

The Method of Galerkin

We will come to grips with the impossibility of satisfying the equations of the model exactly: The solutions will be obtained with an approximate method. There's going to be an error in the balance equation (which we shall call a residual; another appropriate label might be imbalance). Similarly, the natural (force) boundary condition may not be satisfied exactly, and will also produce a residual.

In this book, the approximate solutions are obtained with the Galerkin method. Boris Grigoryevich Galerkin became a teacher of structural mechanics in St. Petersburg Polytechnical Institute in 1908. Among his contemporaries, also active in St. Petersburg, were I. G. Bubnov, A. N. Krylov, and S. P. Timoshenko, all well-known names in various areas of mechanics. In 1915 Galerkin published an article, in which he put forward an idea of an approximate method to solve differential boundary value problems (he was working on plate and shell models at that time). Around that time Bubnov developed similar variational approach, hence this method is also known as the Bubnov-Galerkin method.

2.1 Residual of the balance equation

The balance equation (1.1) may be written in the residual form as

$$P\frac{\partial^2 w}{\partial x^2} + q - \mu\ddot{w} = r_B(x, t) , \qquad (2.1)$$

by moving the inertial force on the other side of the equals sign. The residual r_B is identically zero if w is the exact solution. For an approximate solution, the residual r_B varies from point to point, and from time to time, and is in general nonzero.

Fig. 2.1   Residual that integrates to zero, but is not identically zero.

Checking that the balance residual is identically zero at each point x and each time t does not provide us with anything we can use to talk about approximate solutions: the residual is either zero or it isn't. So how do we measure whether the approximate solution, for which the residual is not zero, is in some sense satisfactory (or not)?

2.2   Integral test of the residual

One possible choice of a quality measure is to integrate the residual over the domain (length of the wire). We could think of the integral

$$\int_0^L r_B(x,t)\,\mathrm{d}x \qquad\qquad (2.2)$$

as a test: if the residual is identically zero, this integral will also come out zero. However, Eq. (2.2) may be zero even when the residual is not identically zero. In other words, if we wanted to prove that the residual corresponded to an exact solution, this would be an incomplete and flawed test. Consider Fig. 2.1: the integral (2.2) is zero, but the residual itself may be very large (for instance, when $r_B = A\sin(2\pi n x/L)$, with $n = 1, 2, ...$).

2.3   Test function

A remedy that addresses this blindness of (2.2) to the shape of the residual may be to use a "window" (test) function $\eta(x)$

$$\int_0^L \eta(x) r_B(x,t)\,\mathrm{d}x \ . \qquad\qquad (2.3)$$

Note that $\eta(x)$ is an arbitrary function. In particular, it could be a function of the shape shown in Fig. 2.2, which is certainly going to give a nonzero value for (2.3) (the hatched area at the bottom). Therefore, it correctly indicates that the residual does not correspond to the exact solution. Equation (2.3) is known as the **weighted residual statement**, because each test function η applies a variable weight to the residual in different parts of the domain. Approximate approaches that start from the weighted residual statement are known as weighted residual methods.

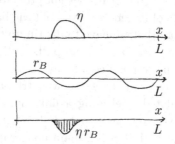

Fig. 2.2   Nonzero residual which is detected in the integral (2.3).

Equation (2.3) is a reliable way of testing the residual, but computationally it seems hardly less difficult than testing the residual at each point of the domain: equation (2.3) needs to be evaluated for an infinite number of functions η in order to make sure there are no bumps in the residual. The job will still take an infinite time.

Let us contemplate a tangible analogy of what we're trying to do in Eq. (2.3). Imagine our job is to hold an inflatable balloon in a box, so that it does not jut out anywhere. Use the fingers of one hand to press down on the balloon, so that the balloon is at the top of the box in the spot where it is being held by the finger. If we put down all five fingers, the situation is as shown on the left in Fig. 2.3. Each of the fingers may be thought of as a single test function η that pushes down the residual in some spots.

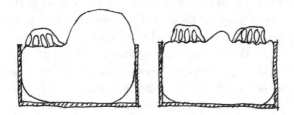

Fig. 2.3   Stuffing a balloon into a box.

Evidently, the balloon bulges out a little bit in between the fingers, and a lot everywhere else. However, we have the option of pressing down on the balloon with the fingers of our other hand, and if we enrol our friends and relatives, and the chance passersby, and distribute the pressing fingers wisely, we will manage to do a better and better job of stuffing the balloon into the box and holding it so that it does not protrude very much. Indeed, with an infinite number of fingers, we can hold the balloon so that it does not protrude at all. Note that we have to distribute the pressing fingers in some sense **densely** and **uniformly**– no parts of the interval $0 \leq x \leq L$ may be left out, since the residual could stay nonzero there.

In this way, we may begin to see how a trial-and-test approximate method may be formulated. Selecting a finite number of suitable functions η_j (fingers), we may be able to control distribution and magnitude of the residual (but, in general, it will remain nonzero). By applying larger numbers of test functions, we will be able to reduce the error in the residual and get a better solution. Also, for each η_j, $j = 1, \ldots, N$, we will make the integral (2.3) vanish

$$\int_0^L \eta_j(x) r_B(x, t) \, \mathrm{d}x = 0 \,, \qquad (2.4)$$

which provides us with the means of calculating N coefficients (numbers) from these N equations.

2.4   Trial function

The task of formulating the approximate solution requires describing the shape of the deflection w. This can be done in a variety of ways, but for reasons that we shall give later, a piecewise linear representation is a good choice. Figure 2.4 illustrates this concept by showing how the shape may be defined by the N coefficients w_j. The attentive reader will at this point fidget: the piecewise linear shape of the deflection curve is not going to allow us to express the second order derivatives $\partial^2 w / \partial x^2$. At the corners, the first derivatives will be discontinuous, and hence the second derivative will be a spike (the so-called Dirac delta function). We can choose either to abandon the piecewise linear shape, and pass a smooth curve through the filled-circle points, or, we could change the rules of the game by getting rid of the second-order derivatives. As we shall presently see, the latter choice is commonly preferred.

In any case, the Eqs. (2.4) may be used to calculate the values of w_j, $j = 1, \ldots, N$. The function that describes the shape of the approximate solution (with the N free parameters) is known as the **trial function.** It describes a possible (candidate, trial) shape of the approximate solution; which becomes *the* solution once the values of the free parameters are known.

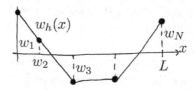

Fig. 2.4 Piecewise linear trial function.

2.5 Manipulation of the residuals

We will seek the approximate solution w to satisfy the balance equation in the residual form (2.4), and we will incorporate the boundary conditions in residual form too. The displacement boundary condition (1.2) will be included in the form of the residual

$$r_w(t) = w(0, t) - \bar{w}_0(t) , \tag{2.5}$$

and the natural boundary condition (1.3) will be included as the residual

$$r_F(t) = -P\frac{\partial w}{\partial x}(L, 0) + F_L . \tag{2.6}$$

Therefore, the approximate solution will be sought from the conditions

$$\begin{aligned} \int_0^L \eta_j(x) r_B(x, t) \, \mathrm{d}x &= 0 , \\ \xi_w r_w(t) &= 0 , \\ \xi_F r_F(t) &= 0 , \end{aligned} \tag{2.7}$$

where ξ_w, ξ_F are test "functions" (actually, in this case just numbers) that are used to test the satisfaction of the boundary conditions. There are no conditions at this point on the trial function w other than smoothness that will guarantee the existence of the integrals in the first equation (2.7).

To reduce the complexity of (2.7), we may immediately realize that at the left-hand end of the wire, we can quite simply *design* the trial function to make the residual (2.5) identically zero. This will impose one more condition on w, and (2.7) may be cast as

$$\int_0^L \eta_j(x) r_B(x,t)\, \mathrm{d}x = 0 \,,$$
$$\xi_F r_F(t) \qquad = 0 \,, \tag{2.8}$$

where $w(0,t) = \bar{w}_0(t)$, and $w(x,t)$ sufficiently smooth in x. Let us remark that Eqs. (2.7) and (2.8) may be interpreted as **orthogonality** conditions.

In this way we managed to reduce the number of residuals, but we will do even better now. By applying integration by parts to the first Eq. (2.8), we will be able to reduce the number of residuals further, and furthermore, we will be able to make it much easier to design a trial function by allowing for less smooth functions.

Substituting for the balance residual, we get three terms

$$\int_0^L \eta_j(x) r_B(x,t)\, \mathrm{d}x =$$
$$\int_0^L \eta_j(x) P \frac{\partial^2 w}{\partial x^2}(x)\, \mathrm{d}x + \int_0^L \eta_j(x) q(x)\, \mathrm{d}x - \int_0^L \eta_j(x) \mu(x) \ddot{w}(x,t)\, \mathrm{d}x \,. \tag{2.9}$$

Integration by parts will not affect the second and third term on the right-hand side, but for the first term we obtain

$$\int_0^L \eta_j P \frac{\partial^2 w}{\partial x^2}\, \mathrm{d}x = \left[\eta_j P \frac{\partial w}{\partial x} \right]_0^L - \int_0^L \frac{\partial \eta_j}{\partial x} P \frac{\partial w}{\partial x}\, \mathrm{d}x \,. \tag{2.10}$$

This identity is fraught with possibilities. If we can figure out what to do with the bracket, we gain the integral on the right in a form that balances derivatives between the test and trial functions. Now for the bracket: Number one, we may recognize part of the bracket in Eq. (2.6). In fact, if we propose to satisfy $r_F = 0$ at the right-hand end of the wire ($x = L$) identically, we may replace $P \frac{\partial w}{\partial x}$ with F_L. That takes care of the force residual (2.6). Number two, at the left-hand end of the wire the value of $P \frac{\partial w}{\partial x}$ is unknown, but we have the option of making η_j vanish at $x = 0$. This will burden all the η_j's with a condition, $\eta_j(x = 0) = 0$, but that is something we can probably afford.

We are in a position to summarize: We have been able to avoid the need to carry the displacement residual (2.5) [eliminated by design of the trial function] and the force residual (2.6) [incorporated into the balance residual– hence, "natural" boundary condition]. Therefore, we will try to find the approximate solution w to satisfy the balance equation in the residual form

$$\eta_j(L)F_L - \int_0^L \frac{\partial \eta_j}{\partial x} P \frac{\partial w}{\partial x} \, dx + \int_0^L \eta_j q \, dx - \int_0^L \eta_j \mu \ddot{w} \, dx = 0, \quad j = 1, \ldots, N \, ,$$

(2.11)

where

$$\eta_j(x = 0) = 0, \quad \eta_j \in C^0, \quad j = 1, \ldots, N \, ,$$
$$w(x = 0, t) = \bar{w}_0(t), \quad w \in C^0,$$
$$w(x, t = 0) \approx \bar{W}(x), \quad \frac{\partial w}{\partial t}(x, t = 0) \approx \bar{V}(x).$$

(2.12)

We write for the trial function $w \in C^0$ and similarly for the test functions. This literally means that the functions are continuous, which is a substitute here for a more precise mathematical statement, but which nevertheless ensures that the integrals in (2.11) exist.

The initial conditions need to be suitably approximated, in general we will not be able to satisfy them exactly (which is why we write $\approx$). Typically, interpolation is used.

2.6 Stiffness and mass matrix

It is time to come back to the choice of the test and trial functions. As advertised in Section 2.4, we have been able to change the requirements on the test and trial function: Their derivatives are now balanced– only the first-order derivatives are needed for either. Therefore, the piecewise linear interpolation function of Fig. 2.4 is now a possibility. However, we can still forge ahead while keeping our options open.

Let us make the assumption that the time is fixed $t = \bar{t}$ ($\bar{t}$ some given number). To describe the trial function, we will resort to a common technique in interpolation which is to write the interpolant as a linear combination of basis functions. Therefore, let us assume that the trial function is written as

$$w(x, \bar{t}) = \sum_{i=1}^N N_i(x) w_i(\bar{t}) \, ,$$

(2.13)

where by $w_i(\bar{t})$ we simply mean that the coefficients of the linear combination w_i are actually functions of time, evaluated at the particular time $\bar{t}$.

Substituting into (2.12), we obtain

$$
\eta_j(L)F_L - \int_0^L \frac{\partial \eta_j}{\partial x} P \sum_{i=1}^N \frac{\partial N_i}{\partial x} w_i(\bar{t}) \, \mathrm{d}x
$$

$$
+ \int_0^L \eta_j q \, \mathrm{d}x - \int_0^L \eta_j \mu \sum_{i=1}^N N_i \ddot{w}_i(\bar{t}) \, \mathrm{d}x = 0, \quad j = 1, \ldots, N \, ,
\tag{2.14}
$$

which may be simplified to

$$
\eta_j(L)F_L - \sum_{i=1}^N \left(\int_0^L \frac{\partial \eta_j}{\partial x} P \frac{\partial N_i}{\partial x} \, \mathrm{d}x \right) w_i(\bar{t})
$$

$$
+ \int_0^L \eta_j q \, \mathrm{d}x - \sum_{i=1}^N \left(\int_0^L \eta_j \mu N_i \, \mathrm{d}x \right) \ddot{w}_i(\bar{t}) = 0, \quad j = 1, \ldots, N \, .
\tag{2.15}
$$

With the definitions

$$
K_{ji} = \int_0^L \frac{\partial \eta_j}{\partial x} P \frac{\partial N_i}{\partial x} \, \mathrm{d}x \, ,
\tag{2.16}
$$

where K_{ji} is usually referred to as the **stiffness matrix**, and

$$
M_{ji} = \int_0^L \eta_j \mu N_i \, \mathrm{d}x \, ,
\tag{2.17}
$$

where M_{ji} is the (consistent) **mass matrix**, we may write (2.15) as

$$
\eta_j(L)F_L - \sum_{i=1}^N K_{ji} w_i(\bar{t}) + \int_0^L \eta_j q \, \mathrm{d}x - \sum_{i=1}^N M_{ji} \ddot{w}_i(\bar{t}) = 0, \quad j = 1, \ldots, N \, ,
\tag{2.18}
$$

where

$$
\eta_j(x = 0) = 0, \quad \eta_j \in C^0, \quad j = 1, \ldots, N
$$

$$
w(x = 0, t) = \bar{w}_0(t), \quad w \in C^0,
\tag{2.19}
$$

$$
w(x, t = 0) \approx \bar{W}(x), \quad \frac{\partial w}{\partial t}(x, t = 0) \approx \bar{V}(x) \, .
$$

The matrix Eq. (2.18) is a system of coupled ordinary differential equations (evaluated at time $\bar{t}$), where the coupling is introduced by the matrices K_{ji} and M_{ji}. The necessary linear algebra would be much more efficient if the two matrices were *symmetric* and *sparse*.

The first property will follow if we take as the test functions η_j the basis functions themselves, $\eta_j \equiv N_j$. The second property may be achieved if the basis functions N_i are nonzero only on a small subset of the interval $0 \le x \le L$.

2.7 Piecewise linear basis functions

Let us recall the piecewise linear approximation proposed for the trial function in Section 2.4. The broken line cannot be represented as a linear combination of linear functions that are all defined on the whole interval $0 \le x \le L$ (only two such functions are linearly independent, and these functions cannot represent the corners in the broken line). Therefore, we have to describe the piecewise linear curve interval by interval.

The interpolant may be written as a linear combination of basis functions. In one dimension, the piecewise linear basis function is called the *hat function*. The six functions that are shown in Fig. 2.5, all are examples of hat functions. For reasons that will be discussed later, we would want the hat functions in a linear combination to be able to reproduce an arbitrary linear function over the whole interval. Because of the way in which we construct the hat functions in Fig. 2.5, this property is automatically available.

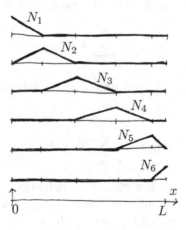

Fig. 2.5 Piecewise linear basis functions.

Let us describe the construction of the piecewise linear basis functions. (In this book, the one-dimensional elements with two nodes at the end points are going to be referred to as L2.) First, the length of the wire is divided into disjoint subintervals. These subintervals are the *finite elements* for the one-dimensional domain. The end-points of the finite elements are called **nodes**. Together, the finite elements and the nodes are known as the **finite element mesh**: see Fig. 2.6 (the element numbers are in the boxes; nodes are indicated by filled circles). Since all basis functions are constructed in the same way, we describe the procedure for the basis function N_3: as shown in the Fig. 2.5, it is nonzero over two elements, 2 and 3; zero everywhere else. To be able to write it down over the two adjacent elements, we have to agree on the value of N_3 at node 3 (i.e. $N_3(x_3)$), which is shared by elements 2 and 3. Choosing $N_3(x_3) = 1$ has certain advantages, which will be introduced momentarily. Using the concept of Lagrange interpolation polynomials, we may write the function N_3 within element 2 as

$$N_3(x) = \frac{x - x_2}{x_3 - x_2}, \quad x_2 \le x \le x_3 ,$$

and within element 3 as

$$N_3(x) = \frac{x - x_4}{x_3 - x_4}, \quad x_3 \le x \le x_4 .$$

All the other functions N_i are expressed analogously. Putting them together in a linear combination for the trial function, we write

$$w(x) = \sum_{i=1}^{N} N_i(x) w_i , \tag{2.20}$$

(for simplicity, we omit the time argument). Evaluating $w(x)$ at the node k, we obtain

$$w(x_k) = \sum_{i=1}^{N} N_i(x_k) w_i ,$$

where the crucial expression is $N_i(x_k)$: by definition, the basis function N_k has value $+1$ at x_k, while all other functions $N_i, i \ne k$ are zero at x_k. This property is usually expressed mathematically as

$$N_i(x_k) = \delta_{ik} , \tag{2.21}$$

where the symbol δ_{ik} is known as the **Kronecker delta**

$$\delta_{ik} = \begin{cases} 1, & \text{if } i = k; \\ 0, & \text{otherwise.} \end{cases}$$

Because of this property, the value of $w(x_k)$ is

$$w(x_k) = \sum_{i=1}^{N} N_i(x_k)w_i = \sum_{i=1}^{N} \delta_{ik}w_i = w_k \ ,$$

and we see that the parameters w_i have the physical meaning of the value of the interpolated function at the node i. The w_i's are usually called the **degrees of freedom**, since, being the control parameters of the trial function, they determine the shape of the actual solution from all the possible shapes of the trial function. They are the objects that our numerical method solves for.

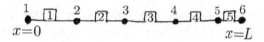

Fig. 2.6   The finite element mesh.

2.8   How are the Galerkin and Finite Element Methods Related

The procedure presented so far may be summarized as follows:

(1) Form residuals r_B and r_F; assume that the trial function will satisfy the essential boundary conditions by design, $w(0,t) = \overline{w}_0(t)$.
(2) Use integration by parts to shift derivatives from the trial function to the test function, subsuming r_F in the weighted residual for the balance equation in the process.
(3) Choose basis functions N_j, and write the trial function as $w(x) = \sum_{i=1}^{N} N_i(x)w_i$.
(4) Choose as the test functions the N_j's.

When the finite element functions (Section 2.7) are chosen as the basis functions N_j, the so-called Galerkin Finite Element Methods (GFEM) result; otherwise we get Galerkin methods that are *not* finite element methods. On the other hand, finite element methods that are not Galerkin methods are quite common in certain applications (fluid mechanics, for instance). This classification is illustrated in Fig. 2.7.

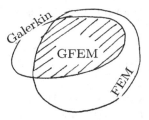

Fig. 2.7　The relationship of Galerkin and finite element methods (FEM). The intersection of the two sets are the Galerkin Finite Element Methods (GFEM).

2.9　Numerical quadrature

In the preceding section we described how to compute the basis function N_3 by visiting the adjacent finite elements on which the function was nonzero. To compute the solution, we actually need a different approach: we need an algorithm to facilitate the numerical evaluation of the integrals in the residual equations. The integrals are calculated element-by-element (the integrands are in general discontinuous from element to element). Therefore, instead of being interested in a single basis function at any point within the mesh, we will strive to calculate the values (and their derivatives) of *all the nonzero basis functions* at a particular point (the quadrature point) within a *single finite element*; see Fig. 2.8. In this element-centric view, we

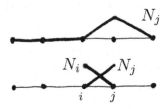

Fig. 2.8　Two different views on how to evaluate basis functions in the finite element mesh: top – compute a single basis function over the whole mesh; bottom – compute all nonzero basis functions over a single element.

would evaluate the functions associated with the nodes at the endpoints of the element. Say for element connecting nodes i, and j, the functions N_i and N_j would be expressed as

$$N_i(x) = \frac{x - x_j}{x_i - x_j}, \quad N_j(x) = \frac{x - x_i}{x_j - x_i} \ . \tag{2.22}$$

It is common practice to develop numerical integration rules on *standard intervals*. Often that will be $-1 \le \xi \le +1$ (line elements in one dimension, quadrilaterals in two dimensions, and bricks in three dimensions all use this interval definition in the so-called tensor-product forms). For instance, Simpson's 1/3 rule is given on this interval as

$$\int_{-1}^{+1} f(\xi)\mathrm{d}\xi \approx \frac{1}{3}f(\xi = -1) + \frac{4}{3}f(\xi = 0) + \frac{1}{3}f(\xi = +1) \, .$$

In general, a numerical quadrature rule would be written on the standard interval $-1 \le \xi \le +1$ as

$$\int_{-1}^{+1} f(\xi)\mathrm{d}\xi \approx \sum_{k=1}^{M} f(\xi_k)W_k \, , \tag{2.23}$$

where ξ_k are the locations of the integration points, and W_k are their weights. Integrating numerically arbitrary functions over arbitrary intervals is then made possible by a map from the standard interval $-1 \le \xi \le +1$ to the arbitrary interval $a \le x \le b$

$$x = \frac{1}{2}(a + b) + \frac{1}{2}(b - a)\xi \, , \tag{2.24}$$

(the first part is the midpoint of the interval $a \le x \le b$, the second part is the departure from the midpoint to either side). Because this map is linear, the relationship between the differentials is constant,

$$\mathrm{d}x = \frac{1}{2}(b - a)\mathrm{d}\xi \, ,$$

where the factor $\frac{1}{2}(b-a)$ is called the **Jacobian determinant**. The Simpson's 1/3 rule is for an arbitrary interval $a \le x \le b$ expressed as

$$\int_{a}^{b} f(x)\mathrm{d}x \approx \frac{1}{2}(b - a)\left[\frac{1}{3}f(a) + \frac{4}{3}f(\frac{1}{2}(a + b)) + \frac{1}{3}f(b)\right] \, .$$

In general, when the map is

$$x = g(\xi), \quad a = g(-1) \, , b = g(+1) \, , \tag{2.25}$$

differentiating both sides with respect to x yields

$$\frac{\mathrm{d}}{\mathrm{d}x}x = 1 = \frac{\mathrm{d}}{\mathrm{d}x}g(\xi) = \frac{\mathrm{d}g(\xi)}{\mathrm{d}\xi}\frac{\mathrm{d}\xi}{\mathrm{d}x} \, ,$$

resulting in

$$\mathrm{d}x = \frac{\mathrm{d}g(\xi)}{\mathrm{d}\xi}\,\mathrm{d}\xi\ ,$$

and the numerical quadrature over an arbitrary interval may be written as

$$\int_a^b f(x)\mathrm{d}x = \int_{-1}^{+1} f(\xi)\frac{\mathrm{d}g(\xi)}{\mathrm{d}\xi}\,\mathrm{d}\xi \approx \sum_{k=1}^M f(\xi_k)\frac{\partial g}{\partial \xi}(\xi_k)W_k\ , \qquad (2.26)$$

where $\dfrac{\partial g}{\partial \xi}(\xi_k)$ is the Jacobian determinant evaluated at the quadrature point ξ_k.

Lets us now look at the calculations for the mass matrix (2.17). The integral will be evaluated over each element individually, and these contributions will be summed together. Therefore, let us consider the integral (2.17) over a single element, connecting nodes i and j

$$\int_{x_i}^{x_j} N_p(x)\mu(x)N_m(x)\ \mathrm{d}x\ , \quad p, m = i, j\ .$$

Note that N_j and N_i are the only two basis functions which are nonzero over the interval $x_i \le x \le x_j$. Clearly, the function $f(x)$ in equation (2.26) is

$$N_p(x)\mu(x)N_m(x)\ , \quad p, m = i, j\ ,$$

which means that we have to express the basis functions in terms of ξ (the mass density $\mu(x)$ is typically constant over an element). Indeed, we have the map (2.24), which upon substitution into (2.22) yields

$$N_i(\xi) = \frac{\xi - 1}{-2}, \quad N_j(\xi) = \frac{\xi + 1}{+2}\ . \qquad (2.27)$$

Basis functions expressed on the standard interval (2.27) are sometimes referred to as being expressed in the ***parametric coordinates***. It is noteworthy that (2.27) are just the Lagrange interpolation polynomials on the standard interval. As we shall see later, writing down the basis functions over a standard shape – a square for general quadrilaterals, a cube for general brick elements, standards triangles or standard tetrahedra for general triangles or general tetrahedra, and so on– is not only convenient, but also highly advisable from the point of view of computer implementation: most of the code for different element shapes and types is then shared, and does not have to be repeated. However, it does mean that we have to express the derivative of the basis functions using a chain rule:

$$\frac{\partial N_i(\xi)}{\partial x} = \frac{\partial N_i(\xi)}{\partial \xi} \frac{\partial \xi}{\partial x} . \tag{2.28}$$

The partial derivative $\frac{\partial \xi}{\partial x}$ is readily obtainable from (2.24), and may be identified as the inverse of the Jacobian.

So, finally we are ready to state the integral of the mass matrix elements

$$\int_{x_i}^{x_j} N_p(x)\mu(x)N_m(x)\, dx \approx \frac{1}{2}(x_j - x_i)\sum_{k=1}^{M} N_p(\xi_k)\mu(\xi_k)N_m(\xi_k)W_k \ ,$$

where the integration rule is as yet undetermined: we could choose any one of the dozens of numerical rules available in the literature. What would the rationale for these choices be? At first sight, of the integrals for the stiffness matrix (2.16) and for the mass matrix (2.17), the latter will require a more accurate numerical quadrature rule: the stiffness matrix involves products of the derivatives of the basis functions, which for linear basis functions are constants; the mass matrix, on the other hand, requires products of the basis functions themselves, which are linear functions of x. Therefore, the stiffness matrix consists of integrals of constants, while the mass matrix elements are integrals of quadratic functions. Interestingly, increased efficiency and even higher accuracy may be occasionally achieved if the mass matrix is *not* integrated exactly. In particular, diagonal (lumped) mass matrices are often used to achieve both benefits in wave propagation problems.

The numerical quadratures that are in common use with polynomial finite elements are the **Gauss rules**. They are well described in a number of textbooks, see for instance the book [Hughes (2000)] and the original references cited there, and we are going to introduce them later.

2.10 Putting it together: system of ODE's

Applying the piecewise linear basis functions derived in Section 2.7 to Eq. (2.19) is straightforward. The unknown degrees of freedom are $w_2(t)$, $w_3(t)$, ..., $w_N(t)$; the function $w_1(t)$ is given by the boundary conditions: the condition $w(x = 0, t) = \bar{w}_0(t)$ becomes, as a result of the Kronecker delta property, $w_1(t) = \bar{w}_0(t)$.

The stiffness and mass matrices will be symmetric, tri-diagonal, i.e.

$$K_{ji} = \int_0^L \frac{\partial N_j}{\partial x} P \frac{\partial N_i}{\partial x} \, dx \begin{cases} \neq 0, \text{ if } |i - j| \leq 1; \\ = 0 \text{ otherwise,} \end{cases} \tag{2.29}$$

and

$$M_{ji} = \int_0^L N_j \mu N_i \, dx \begin{cases} \neq 0, \text{ if } |i - j| \leq 1; \\ = 0 \text{ otherwise.} \end{cases} \tag{2.30}$$

Equation (2.15) is trivially modified to read

$$N_j(L)F_L - \sum_{i=1}^N K_{ji} w_i(t) + \int_0^L N_j q \, dx - \sum_{i=1}^N M_{ji} \ddot{w}_i(t) = 0, \quad j = 2, \ldots, N \ .$$

$$\tag{2.31}$$

Note that j runs from 2 to N (as explained above), but i ranges over all nodes, i.e. also the first degree of freedom is included, even though it is determined from the boundary condition. In effect, nonzero displacement $w_1(t)$ generates an external force with the jth component

$$-K_{j1} w_1(t) - M_{j1} \ddot{w}_1(t) \ .$$

The second order differential equations (2.18) may be integrated for instance by converting them to first order form and using an off-the-shelf Matlab integrator. However, because of their special form, there are excellent custom-tailored algorithms for this purpose: for example the Newmark explicit algorithm.

Exercises

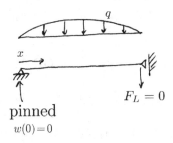

Fig. 2.9   Prestressed wire.

(1) Consider a cable loaded statically by a sinusoidal distribution of transverse load

$$q = \bar{q}\sin(\frac{\pi x}{L}),$$

with $\bar{q} = 25$, $L = 100$. The prestressing force is

$$P = 100\bar{q}L.$$

The left-hand end is pinned, and there's no force applied at the right-hand end. Compute the approximate solution for the deflection of the wire from the Galerkin formulation. Consider a one-term approximation with the test function $\eta_1 = x$, and the basis function $N_1 = \sin(\frac{\pi x}{2L})$.

    (a) Compute the multiplier w_1 and plot the solution. Compare with the analytical solution.

    (b) Compute the force residual r_F at $x = L$.

(2) Same data as in assignment (1). Use test function $\eta_1 = N_1$.

    (a) Compute the multiplier w_1 and plot the solution.

    (b) Compute the force residual r_F at $x = L$.

(3) Same data as in assignment (1). Use test function $\eta_1 = N_1 = x$.

    (a) Compute the multiplier w_1 and plot the solution.

    (b) Compute the force residual r_F at $x = L$.

(4) Same data as in assignment (1). Use test function $\eta_1 = N_1 = \sin(\frac{\pi x}{L})$.

    (a) Compute the multiplier w_1 and plot the solution.

    (b) Compute the force residual r_F at $x = L$.

(5) Same data as in assignment (1). Use test function $\eta_1 = N_1 = \sin(\frac{\pi x}{L})$.

    (a) Compute the multiplier w_1 and plot the solution.

    (b) Compute the balance residual r_B for the approximate solution,

$$r_B = P\frac{\partial^2 w_{\text{approx}}}{\partial x^2} + q.$$

    and plot it.

(6) Same data as in assignment (1). Use test function $\eta_1 = N_1 = \sin(\frac{\pi}{2}\frac{x}{L})$.

    (a) Compute the multiplier w_1 and plot the solution.

(b) Compute the balance residual r_B for the approximate solution,

$$r_B = P \frac{\partial^2 w_{\text{approx}}}{\partial x^2} + q \ .$$

and plot it.

(7) Same data as in assignment (1), except the boundary condition at the left-hand end point is

$$w(0) = \overline{w}_0 \ ,$$

where $\overline{w}_0 = -0.13$. Use test function $\eta_1 = N_1 = \sin(\frac{\pi}{2} \frac{x}{L})$.

(a) How does the approximate deflection w_{approx} need to be defined in order to satisfy the boundary condition? Compute the multiplier w_1 and plot the solution.

(8) Figure 2.10: Consider a cable loaded statically by a uniform distribution of transverse load

$$q = \overline{q} \ ,$$

with $\overline{q} = 1.5$, $L = 100$. The prestressing force is

$$P = 100\overline{q}L \ .$$

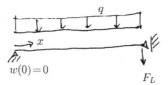

Fig. 2.10   Prestressed wire with uniform transverse load and a force.

The left-hand end is pinned, and there's a force $F_L = -75$ applied at the right-hand end. Compute the approximate solution for the deflection of the wire from the Galerkin formulation (2.18), with the conditions on test and trial function (2.12). Consider a one-term approximation with the basis function $N_1 = \sin(\frac{\pi x}{L})$, and the test function $\eta_1 = N_1$.

(a) Compute the approximate deflection w_{approx}.

(b) Compute the boundary force residual r_F.

(9) Problem definition as in (8). It will be useful to use the symbolic toolbox that comes with Matlab, or perhaps some other computer algebra tool, to facilitate the calculations.

(a) Compute the approximate deflection w_{approx} and plot it to compare with the analytical solution. Consider a two-term approximation with the basis functions $N_1 = \sin(\frac{\pi}{2}\frac{x}{L})$, $N_2 = \sin(\pi\frac{x}{L})$ and the test functions $\eta_1 = x$, and $\eta_2 = x^2$.

(b) Compute the approximate deflection w_{approx} and plot it to compare with the analytical solution. Consider a two-term approximation with the basis functions $N_1 = x$, and $N_2 = x^2$ and the test functions $\eta_1 = \sin(\frac{\pi}{2}\frac{x}{L})$, $\eta_2 = \sin(\pi\frac{x}{L})$.

(c) Compute the approximate deflection w_{approx} and plot it to compare with the analytical solution. Consider a two-term approximation with the basis functions $N_1 = x$, and $N_2 = x^2$ and the test functions $\eta_1 = N_1$, $\eta_2 = N_2$.

Chapter 3

Statics and Dynamics Examples for the Wire Model

In this chapter we will introduce a finite element library (or toolbox, if you prefer), SOFEA[1]. It will be used to produce finite element solutions using the results of the previous chapter for the Galerkin method. SOFEA is written for Matlab, and its design is based on the object oriented support in Matlab. In particular, all the methods and algorithms are present in the library as methods defined for classes.

The toolbox is available for download at

http://hogwarts.ucsd.edu/~pkrysl/sofea .

Matlab needs to know where to find the various functions and classes. Therefore, please be sure to run the initialization script sofea_init. (For the Windows environment, you may start Matlab and run sofea_init by double-clicking the start.bat batch file.) For details, please refer to the README file.

Matlab is adequately described in the online documentation [Matlab]. The "Getting started" tutorial is especially helpful for beginners. Also, Moler has published an accessible book about numerical analysis and Matlab [Moler], including a brief tutorial on the use of Matlab. The book can also serve as reference for any numerical analysis issues that are touched upon in this book.

The Matlab online documentation [Matlab] has also a well-written discussion of the Matlab object-oriented features (under "Matlab"/"Programming"/"Classes and Objects"). The SOFEA toolbox employs only the most basic class-based facilities of the language, and the design of the toolbox is quite simple.

---

[1] SOFEA is © 2005, Petr Krysl

3.1   Statics

When the inertial forces may be neglected in the balance equation, we have the case of statics (static equilibrium). The Galerkin formulation omits the terms with the accelerations, and reads

$$N_j(L)F_L - \sum_{i=1}^{N} K_{ji}w_i + \int_0^L N_j q \,\mathrm{d}x = 0, \quad j = 2,\ldots,N\ , \qquad (3.1)$$

which may be arranged in matrix form as

$$\boldsymbol{Kd} = \boldsymbol{L}\ , \qquad (3.2)$$

where $\boldsymbol{K}$ is a square $(N-1) \times (N-1)$ matrix collecting K_{ji}, $i,j = 2,\ldots,N$. The column matrix $\boldsymbol{d}$ collects the degrees of freedom $d_k = w_{k+1}$, $k = 1,\ldots,N-1$. The column matrix $\boldsymbol{L}$ is the load vector, with components

$$L_k = N_{k+1}(L)F_L - K_{k+1,1}w_1 + \int_0^L N_{k+1}q \,\mathrm{d}x = 0, \quad k = 1,\ldots,N-1\ .$$
$$(3.3)$$

3.2   Statics: uniform load

In this example, we assume the transverse load q is uniform, and the transverse force is absent, $F_L = 0$. The Matlab script implementing the solution is w1 [2]. It starts with the definition of the variables.

```
0001 disp('Taut wire: example 1-- statics, uniform load');
0002 L=6;
0003 P=4;
0004 q =-0.1;
```

Next, the mesh is defined: an array of nodes is created, with node 1 at $x = 0$ and so on. The function fenode is the constructor of the class fenode, and the attributes are being passed as fields of a struct, as pairs "name, value" (for instance, 'id',j): this approach is uniformly adopted for all constructors in SOFEA. The array gcells collects finite elements of the type gcell_L2. The attribute conn is the connectivity: the numbers of nodes that are connected by the element. The finite elements are referred to as *geometric cells*. The main reason is that the finite elements have rather

---

[2]Folder: SOFEA/examples/taut_wire

limited responsibilities in SOFEA, namely calculation of the basis functions
(and their derivatives), and drawing of the shape of the cell are essentially
all that is required. They represent a piece of the computational domain,
a geometrical cell.

```
0005 n=2; % number of elements
0006 % Mesh
0007 x=0;
0008 fens=[];
0009 for j= 1:n+1
0010     fens=[fens fenode(struct ('id',j,'xyz',[x]));];
0011     x = x+(L/n);
0012 end
0013 gcells = [];
0014 for j= 1:n
0015     gcells=[gcells gcell_L2(struct('id',j,
                        'conn',[j j+1]))];
0016 end
```

The operations on the mesh that reflect the particular problem that is
being solved are encapsulated in a class descended from the finite element block class, feblock. In particular, the prestressed wire stiffness
and mass matrix, the effect of the distributed load, q, and the effect of the
nonzero displacement at $x = 0$, are computed by the methods of the class
feblock_defor_taut_wire. Note that Simpson's 1/3 rule is being used for
the numerical quadrature. The finite element block consists of all the finite
elements in the array gcells.

```
0018 feb=feblock_defor_taut_wire(struct('mater',mater_defor,...
0019     'gcells',gcells,...
0020     'integration_rule',simpson_1_3_rule,...
0021     'P',P));
```

The quantities that are interpolated on the finite element mesh, such as
the transverse displacement of the wire, w, or the geometry of the mesh, are
represented in SOFEA as instances of the class field. The field geom records
the geometry of the mesh. The constructor on line 0023 retrieves the nodal
coordinates from the array of the nodes. The dimension of the field is 1
because each degree of freedom is just a single displacement. On line 0025
we define the field of the transverse displacements, w. For convenience, it
is defined by cloning the geom field, and then zeroing out all the degrees of

freedom.

```
0023 geom = field(struct ('name',
        ['geom'], 'dim', 1, 'fens',fens));
0025 w    = 0*clone(geom,'w');
```

Next, the displacement (essential) boundary conditions are defined, and applied to the displacement field. The method `set_ebc` only records which components of the degrees of freedom, at which node, are being prescribed (or released), and to which value are they being prescribed. The method `apply_ebc` is then used to transfer this information to the actual degrees of freedom. Finally, the method `numbereqns` numbers the degrees of freedom that are not being prescribed, effectively assigning each one a global equation number.

An important remark should be made here: Lines 0028, 0029, and 0030 illustrate a design feature of Matlab, where all arguments are passed by value. Therefore, no matter what we do with the arguments inside the functions, the variables that were passed by the caller into those functions do not change at all. The functions work with copies, not the actual variables that the caller passed. If the caller wishes to change the variables, the method must return the changed value, and the caller must assign this value. Example: on line 0031 the method `numbereqns` will number the equations in a copy of the field w, and then will return the copy. Since we assign back to the field w, the computed numbering of the equations will be now available in w; if we did not assign back to w, all the work done by the method `numbereqns` would be forgotten.

```
0027 fenids=[1]; prescribed=[1]; component=[1]; val=0;
0028 w    = set_ebc(w, fenids, prescribed, component, val);
0029 w    = apply_ebc (w);
0030 % Number equations
0031 w    = numbereqns (w);
```

Next we create the global system of equations. The stiffness matrix is an object, K, which gets created in line 0033, and initialized to represent a dense neqns×neqns matrix. In line 0034, the stiffness matrices calculated for each finite element by the block feb are assembled into the global matrix K (class dense_sysmat). The number of equations is being retrieved from the displacement field (the global equation numbers have been placed there above) using the get method. The get method is available for all SOFEA objects, and just typing w at the command line produces a list of all the properties that can be obtained from the object. A great way to explore SOFEA objects is the graphical user interface of the object browser, OBgui (part of SOFEA). Its operation is supported by the output of get(w) (notice that only the object itself is passed as argument): using get in this way returns a cell array, with the name and the description of each attribute.

```
0033 K = start (dense_sysmat, get(w, 'neqns'));
0034 K = assemble (K, stiffness(feb, geom, w));
```

The load q is in this case represented by the body_load class. The global load object sysvec is assembled from element load vectors computed by the finite element block.

```
0036 bl = body_load(struct ('magn',inline(num2str(q))));
0037 F = start (sysvec, get(w, 'neqns'));
0038 F = assemble (F, body_loads(feb, geom, w, bl));
```

Finally, the global stiffness object is asked to produce the actual stiffness array, and the global load object is asked to supply the actual load vector array. The standard backslash Matlab operator than computes the solution, which is then stored in the proper places in the displacement field w. The method scatter_sysvec distributes the system vector (the solution of the system of linear equations) to the proper degrees of freedom, and we should note that again the result is assigned to w.

```
0040 w = scatter_sysvec(w, get(K,'mat')\get(F,'vec'));
```

A graphical presentation is generated by plotting the linear coordinate (gathered from the geometry field geom) versus the linear interpolation of the approximate displacements (gathered from w). The analytical solution is also plotted (line 0043). It is noteworthy that the approximate solution interpolates the analytical solution, but such behavior is peculiar to the one-dimensional problem and does not occur for the multi-dimensional models discussed later in the book.

```
0042 xs= (0:0.01:L);
0043 plot (xs, -q/P*xs.*(xs/2-L),'r-','linewidth', 3);
0044 hold on
0045 plot (gather (geom, (1:n+ 1),'values'), ...
0046      gather (w, (1:n+ 1),'values'),'bo-','linewidth', 3);
0047 figure (gcf)
```

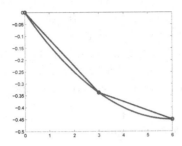

Fig. 3.1   The displacements of the taut wire; the exact solution curve, and the approx-imation by the broken line.

3.3   Free vibration

If we remove all external loads, and prescribe homogeneous (zero) displacements, the Galerkin formulation reads

$$- \sum_{i=2}^{N} K_{ji} w_i(t) - \sum_{i=2}^{N} M_{ji} \ddot{w}_i(t) = 0, \quad j = 2, \dots, N , \qquad (3.4)$$

which may be arranged in matrix form as

$$\boldsymbol{K}\boldsymbol{w} + \boldsymbol{M}\ddot{\boldsymbol{w}} = \boldsymbol{0} , \qquad (3.5)$$

where $\boldsymbol{K}$ is a square $(N-1) \times (N-1)$ matrix collecting K_{ji}, $i, j = 2, \dots, N$, and analogously for the mass matrix. The column matrix $\boldsymbol{w}$ collects the degrees of freedom $w_i(t)$. Equations (3.5) represent the so-called *free vibration* response. The solution is sought in the form $\boldsymbol{w}(t) = \boldsymbol{\phi} \exp(\mathrm{i}\omega t)$, which leads to the *generalized eigenvalue problem*

$$\boldsymbol{K}\boldsymbol{\phi} - \omega^2 \boldsymbol{M}\boldsymbol{\phi} = \boldsymbol{0} , \qquad (3.6)$$

where ω is the circular frequency, and $\boldsymbol{\phi}$ is the eigenmode.

As an example, the present model will be used to calculate the first five natural frequencies of a simply-supported taut string of constant mass density. The Matlab script `wvib`[3] obtains the solution for a series of progressively finer and finer meshes (from 8 elements to 1024 elements). The mass matrix is either computed from the formula (2.17) (this is the so-called *consistent mass matrix*), or it is diagonalized (lumped) by assigning each node half the mass of the adjacent finite elements

$$M_{ji} = \mu h \delta_{ji} , \qquad (3.7)$$

---

[3] Folder: `SOFEA/examples/taut_wire`

which results in the so-called *lumped mass matrix*; we assumed μ uniform, and h to be the length of each finite element.

The analytical formula for the natural frequency [Graff (1991)]

$$\omega^2 = \frac{P}{\mu}\left(\frac{n\pi}{L}\right)^2 \; , \quad n = 1, 2, 3, \ldots$$

produces reference values which are compared with the frequencies solved for from (3.6). The results are summarized in Fig. 3.2. The progressive reduction of the error due to the use of more and more elements is called *convergence*. It may be observed that the two mass matrix formulations lead to convergence from different sides: consistent mass matrix overestimates the natural frequencies, while the lumped mass matrix underestimates them. As to the accuracy: clearly, for quite reasonable engineering tolerance of 5% error, it takes 32 elements for either formulation of the mass matrix to compute all five natural frequencies within the tolerance.

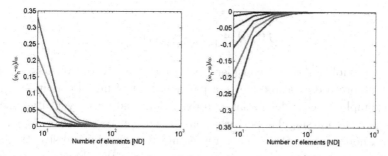

Fig. 3.2 Convergence of the first five natural frequencies of vibration; left: consistent mass matrix, right: lumped mass matrix. Vertical axis: normalized error $(\omega_h - \omega)/\omega$.

3.4 Integration of transient motion

In addition to analytical approaches, the system of ordinary differential equations (2.18) may be integrated numerically. In this section, we will apply an off-the-shelf Matlab integrator.

Matlab integrators work with a system of first order differential equations. Therefore, (2.18) will be first converted to this form. For convenience, Eq. (2.18) will be cast in matrix form

$$\boldsymbol{M\ddot{w}} + \boldsymbol{Kw} = \boldsymbol{F} \; . \tag{3.8}$$

The stiffness matrix entries are defined in (2.16). The mass matrix may be either of the consistent variety (2.17), or a special-purpose matrix such as the lumped mass (3.7). Note that w is the vector function of time-dependent deflections.

Defining $v = \dot{w}$, the first order system may be written as

$$\begin{bmatrix} 1 & 0 \\ 0 & M \end{bmatrix} \begin{bmatrix} \dot{w} \\ \dot{v} \end{bmatrix} = \begin{bmatrix} v \\ -Kw + F \end{bmatrix}, \qquad (3.9)$$

which is in the "mass matrix" form

$$\widetilde{M}\dot{y} = f(t, y) . \qquad (3.10)$$

Here

$$y = \begin{bmatrix} w \\ v \end{bmatrix} ,$$

and

$$f(t, y) = \begin{bmatrix} v \\ -Kw + F \end{bmatrix} .$$

In what follows we will consider a particular example of transient vibrations: unforced oscillation due to a particular set of initial conditions. In this example we consider initial zero deflection, and initial velocity which corresponds to a singular perturbation: a single node at the midspan is given a nonzero velocity. This is representative of a physical situation in which a stationary taut string is rapped sharply with a hammer. Waves propagate away from the midpoint, hit the fixed end-points, and complex interference pattern develops (see Fig. 3.3).

3.4.1   *Using built-in Matlab solver*

The Matlab script wtransient1[4] computes the solution to the transient vibration problem using a built-in Matlab solver. As expected, the initial conditions need to be supplied.

```
0024    w0 = gather_sysvec (w);
0025    v0 = w0; v0(round(n/2)) = 1;
```

Further, the system matrices are computed.

---

[4]Folder: SOFEA/examples/taut_wire

```
0027    K = start (dense_sysmat, get(w, 'neqns'));
0028    K = assemble (K, stiffness(feb, geom, w));
0029    Kmat =get(K,'mat');
0031    M = start (dense_sysmat, get(w, 'neqns'));
0032    % Assemble the lumped mass matrix
0033    M = assemble(M, elemat( struct('eqnums',...
               (1:n-1),'mat', eye(n-1) * mu*L/n)));
0034    Mmat=get(M,'mat');
0035    Mattilda= [eye(get(w, 'neqns')),
               zeros(get(w, 'neqns'));...
0036           zeros(get(w, 'neqns')),Mmat];
```

Matlab provides a suite of several ODE solvers: ode23 is an implementation of an explicit Runge-Kutta (2,3) pair of Bogacki and Shampine [Bogacki, Shampine (1989)]. These solvers use a standard argument list, with one of the arguments being a function handle to the function that evaluates the right-hand side of (3.10). In our case, this function is written as

```
0038    function rhs =f(t,y)
0039        neq=length(y)/2;
0040        w=y(1:neq); v=y(neq+1:end);
0041        rhs= [v;-Kmat*w];
0042    end
```

The invocation of the solver is unremarkable, except that we use odeset to supply the mass matrix, $\widetilde{M}$. The output arguments collect an array of output times, and an array of calculated deflections at all internal nodes.

```
0043    [ts,ys]=ode23(@f,[0, 1500],[w0;v0],...
               odeset('Mass',Mattilda,'RelTol',1e-3));
```

3.4.2 *Using the Trapezoidal integrator*

The Runge-Kutta of the previous section seems to be doing an adequate job. However, integrating the second order equations by first converting them to a first order system is sub-optimal: Firstly, the dimension of the ODE system is doubled. Secondly, the Runge-Kutta ode23 integrator is not actually doing a very good job as it fails to maintain the balance of energy.

Energy conservation is a very important indicator of the quality of the

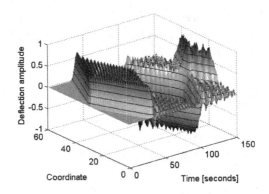

Fig. 3.3   Amplitude of the vertical deflection as a function of the coordinate x and the time.

numerical solution, since energy *should* be conserved in the exact solution to this particular problem. For the taut string problem, the total energy may be defined as

$$TE = \frac{1}{2} \left(\dot{w} \cdot M \cdot \dot{w} + w \cdot K \cdot w \right) . \tag{3.11}$$

It may be observed that the solution produced by the **ode23** integrator does not conserve this quantity (refer to Fig. 3.4), and that provides motivation for the development of algorithms specialized to mechanical systems.

The algorithm we are going to develop next is a special form of the well-known Newmark integrator which is very popular and well-respected in the computational mechanics community for a number of reasons [Hughes (2000)]. The starting point is the first order system (3.10). Integrating in time

$$\int_{t_0}^{t} \widetilde{M} \dot{y} \, \mathrm{d}\tau = \int_{t_0}^{t} f(\tau, y) \, \mathrm{d}\tau , \tag{3.12}$$

yields

$$\widetilde{M}(y_t - y_{t_0}) = \int_{t_0}^{t} f(\tau, y) \, \mathrm{d}\tau . \tag{3.13}$$

The right-hand side integral may be approximated using the trapezoidal rule, which leads to the algorithm

$$\widetilde{M}(y_t - y_{t_0}) = \frac{t - t_0}{2} \left(f(t, y_t) + f(t_0, y_{t_0}) \right) . \tag{3.14}$$

Instead of using the first-order form, we may multiply through in (3.14), obtaining a coupled system

$$\boldsymbol{w}_t = \boldsymbol{w}_{t_0} + \frac{t - t_0}{2}\left(\boldsymbol{v}_t + \boldsymbol{v}_{t_0}\right)$$

$$\boldsymbol{M}\boldsymbol{v}_t = \boldsymbol{M}\boldsymbol{v}_{t_0} - \frac{t - t_0}{2}\boldsymbol{K}\left(\boldsymbol{w}_t + \boldsymbol{w}_{t_0}\right) + \frac{t - t_0}{2}\left(\boldsymbol{F}_t + \boldsymbol{F}_{t_0}\right) , \quad (3.15)$$

which may be marched forward by substituting the first equation into the second, solving for $\boldsymbol{v}_t$, and then updating $\boldsymbol{w}_t$ from the first equation. The integrator obtained in this way is a special case of the general Newmark integrator [Hughes (2000)]. The Newmark algorithm has two free parameters, and for the special choice $\gamma = 1/2$ and $\beta = 1/4$ (the so-called average acceleration method) one obtains the trapezoidal integrator (3.15).

The Matlab script wtransient2[5] computes the solution to the transient vibration problem using a trapezoidal integrator (Newmark average-acceleration integrator). Note that the integrator algorithm is implemented as a nested function, which has access to variables defined in the enclosing environment, the stiffness matrix Kmat and the mass matrix Mmat.

```
0033    Mmat=get(M,'mat');
0034    % Solve
0035    function [ts,ys] = trapezoidal(nsteps,tspan,w0,v0)
0036        ts= zeros(nsteps, 1);
0037        nu =length(w0);
0038        ys = zeros(2*nu,nsteps);
0039        dt = (tspan(2)-tspan(1))/nsteps;
0040        t =tspan(1);
0041        for i=1:nsteps
0042            ts(i) =t;
0043            ys(1:nu,i) =w0; ys(nu+1:end,i) =v0;
0044            v1=(Mmat+((dt/2)^2)*Kmat)\...
                    ((Mmat-((dt/2)^2)*Kmat)*v0-dt*Kmat*w0);
0045            w1=w0+dt/2*(v0+v1);
0046            w0=w1;v0=v1;
0047            t=t+dt;
0048        end
0049        ys=ys';
0050    end
0051    % Now call the integrator
```

---

[5]Folder: SOFEA/examples/taut_wire

0052   [ts,ys]=trapezoidal(3000,[0, 1500],w0,v0);

Figure 3.4 compares the computed total energy for the two integrators introduced in this section. Evidently, the nominally less accurate trapezoidal (Newmark average acceleration) integrator does a much better job than the Runge-Kutta ode23 integrator.

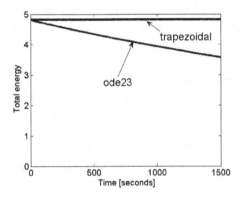

Fig. 3.4   Total energy obtained with two different time integrators.

Exercises

(1) Consider a cable pinned at the left-hand end, and free to slide at the right-hand end (see Figure 3.5). The mass density is $\mu = 0.61$. The prestressing force is $P = 150L$, and the length is $L = 100$.

Fig. 3.5   Prestressed wire schematic for the vibrations study.

Compute the approximate solution for the first natural frequency of the wire. Consider a one-term approximation with the basis function $N_1 = \sin(\pi \frac{x}{L})$, and the test function $\eta_1 = N_1$.

(a) Compute the first natural frequency, and plot the mode shape. Compare with the analytical solution.

(2) Data as in assignment (1). Compute the approximate solution for the first natural frequency of the wire. Consider a two-term approximation with the basis functions $N_1 = x$, $N_2 = x^2$, and the test functions $\eta_j = N_j$.

    (a) Compute the first natural frequency, and plot the mode shape. Compare with the analytical solution.

(3) Data as in assignment (1). Compute the approximate solution for the first two natural frequencies of the wire. Consider a two-term approximation with the basis functions $N_1 = \sin(\frac{\pi}{2}\frac{x}{L})$, $N_2 = \sin(\pi\frac{x}{L})$, and the test functions $\eta_j = N_j$.

    (a) Compute the first and second natural frequency, and plot the mode shape. Compare with the analytical solution.

(4) Refer to the Matlab script `wvib`[6] for the definition of a free-vibration problem of a simply supported cable. Depending on the definition of the mass matrix (consistent – exactly integrated, lumped – integration with one-point quadrature at the nodes), the computed natural frequencies either over-predict or under-predict. Therefore, it seems natural to try mixing together mass matrices obtained with different formulations. Consider the blending equation

$$M = \theta M_c + (1 - \theta)M_l \,,$$

where M_c is the consistent mass matrix of the cable, and M_l is the lumped mass matrix.

    (a) Modify the script `wvib` to compute the first five natural frequencies based on the blended mass matrix M, and determine approximately an "optimal" value of the coefficient θ to attain the best accuracy for the first five frequencies simultaneously. Use meshes with $8, 16, 32, 64, 128$ elements. Turn in the plot of the convergence of the natural frequencies for the (nearly) optimal value of θ, and the printout of the modified script.

(5) Consider a cable which is attached to sliding supports at both ends. (see Figure 3.6). The mass density is $\mu = 80$. The prestressing force is $P = 0.9$, and the length is $L = 7$.

---

[6] Folder: `SOFEA/examples/taut_wire`

Fig. 3.6   Prestressed wire with sliding boundary conditions at both ends.

(a) Compute the first two natural frequencies assuming a single-element finite element mesh (type L2). What is their interpretation? This is a paper and pencil exercise.

(b) Modify the Matlab script wvib[7] to solve for the first two natural frequencies using an eight element mesh; compute with a consistent mass matrix. [Hint: The Matlab function eigs requires certain rank properties from the input matrices: one possible way of increasing the rank of a matrix is to add to it a multiple of the identity matrix.] Compare with the analytical solution, and turn in a plot of the two modes shapes, and a printout of the modified script. Label your modifications in the printout, and give a brief justification for each.

(6) Prove that the trapezoidal integrator conserves exactly the total energy (3.11) in the absence of external loads. [Hint: Write down the total energy at two subsequent times, and compare these expressions.] This is a paper and pencil exercise.

---

[7]Folder: SOFEA/examples/taut_wire

Chapter 4

Boundary Conditions for the Model of a Taut Wire

In this chapter we will explore the effect the boundary conditions have on the solution, both its existence and its computability with the Galerkin model.

We will consider only statics, so that the balance equation (1.1) drops the inertial term

$$P\frac{\partial^2 w}{\partial x^2} + q = 0 \ , \tag{4.1}$$

and furthermore we will assume that the transverse load q is a constant. With the definition $k = -q/P$, the task is to integrate

$$\frac{\partial^2 w}{\partial x^2} = k = \text{constant} \ , \tag{4.2}$$

which is easily accomplished as

$$w(x) = k\frac{x^2}{2} + Cx + D \ , \tag{4.3}$$

where C and D are integration constants to be determined from the boundary condition.

For the model of the wire the boundary of the domain is composed of two disjoint sets, each consisting of a single point at either end. As already mentioned (Section 1.4), one condition on the boundary can be used to determine the solution. Since the balance equation is of second order, and the unknown is a single function, a single boundary condition is all that is needed (with caveats, however). In one dimension, a simple explanation is that a second order equation needs two integration constants; we will talk about this issue in higher dimensional domains when we deal with the heat

41

conduction equation and also, in a lot more detail, in the part dedicated to the elasticity model.

The boundary condition may be of two distinct types: either prescribed deflection (an essential boundary condition), or prescribed slope (derivative of the deflection, which is a natural boundary condition). Since we may prescribe only one boundary condition, but the boundary consists of two disjoint sets, we may in fact prescribe one or the other type at either of the two endpoints. Both existence and uniqueness of the solution depend on which type of boundary condition is applied, and the various possibilities are discussed in this chapter.

4.1   Mixed essential and natural boundary conditions

This case of boundary condition has been treated in the previous few chapters. The natural condition may be expressed in terms of the end-point transverse forces. At $x = L$ the relation is Eq. (1.3), and it may be derived in the same way at $x = 0$ as

$$P\frac{\partial w}{\partial x}(0) + F_0 = 0 \,. \tag{4.4}$$

The Galerkin algebraic equations for an essential boundary condition at $x = 0$ and a natural boundary condition at $x = L$ are given by Eq. (2.18) (omitting the accelerations). When the points of application of these boundary conditions are switched, the changes are limited to the boundary term and the conditions for the test and trial functions only

$$N_j(0)F_0 - \sum_{i=1}^{N} K_{ji}w_i + \int_0^L N_j q \, dx = 0, \quad j = 1,\dots,N-1 \,, \tag{4.5}$$

where

$$N_j(x = L) = 0, \quad N_j \in C^0, \quad j = 1,\dots,N-1$$
$$w_N = \bar{w}_L \,. \tag{4.6}$$

After the solution for the deflection has been obtained, the slope at the end with the essential boundary condition may be computed, yielding the associated reaction force. For instance, for the problem (4.5), (4.6), the analytical solution of (4.3), with the boundary conditions

$$\frac{\partial w}{\partial x}(0) = -\frac{F_0}{P} = \bar{w}_0', \quad w(L) = \bar{w}_L \,,$$

is

$$w(x) = \frac{k}{2}(x^2 - L^2) + \bar{w}'_0(x - L) + \bar{w}_L .$$

This yields the reaction force $F_L = P(kL + \bar{w}'_0)$.

4.2 Essential boundary conditions only

The boundary conditions are in this case the deflections given at both ends,

$$w(0) = \bar{w}_0, \quad w(L) = \bar{w}_L ,$$

which may be substituted into (4.3) for $x = 0$ and $x = L$ to yield two equations for C and D. Upon substitution, the slopes at the end points are available as

$$\frac{\partial w}{\partial x}(0) = \frac{\bar{w}_L - \bar{w}_0}{L} - k\frac{L}{2}, \quad \frac{\partial w}{\partial x}(L) = \frac{\bar{w}_L - \bar{w}_0}{L} + k\frac{L}{2} .$$

From the slopes, the forces F_0 and F_L are available from (4.4) and (1.3). The physical meaning of these two forces is that of reactions at the supports. The algebraic equation for this type of boundary conditions becomes

$$-\sum_{i=1}^{N} K_{ji}w_i + \int_0^L N_j q \, dx = 0, \quad j = 2, \ldots, N-1 , \qquad (4.7)$$

where

$$N_j(x = 0) = 0, N_j(x = L) = 0, \quad N_j \in C^0, \quad j = 1, \ldots, N-1,$$
$$w_1 = \bar{w}_0, \quad w_N = \bar{w}_L . \qquad (4.8)$$

Note that the degrees of freedom to be computed are only at the interior nodes.

4.3 Natural boundary conditions only

The boundary conditions are in this case the slopes given at both ends,

$$\frac{\partial w}{\partial x}(0) = \bar{w}'_0, \quad \frac{\partial w}{\partial x}(L) = \bar{w}'_L,$$

Integrating (4.2) once yields for the slope

$$\frac{\partial w}{\partial x}(x) = kx + C ,$$

which may be used in conjunction with the boundary conditions to express the constant C as either of the two expressions

$$C = \bar{w}_0' \ , \quad \text{or} \quad C = \bar{w}_L' - kL \ .$$

Clearly, this is only possible if

$$\bar{w}_0' = \bar{w}_L' - kL \ ,$$

which may be interpreted as a condition under which a solution exists: if the two slopes are linked by the previous equation, solution exists; otherwise no solution exists, because the boundary conditions are contradictory.

If the solution exists, and the two slopes are not independent, the boundary conditions are really not going to be sufficient to determine two constants of integration, but only one. Correspondingly, the deflection of the wire is then

$$w(x) = k\frac{x^2}{2} + \bar{w}_0' x + D \ ,$$

where the constant D remains undetermined.

An initial boundary value problem with natural boundary conditions only is called a **pure-traction problem**, or Neumann problem. We could see that the solution then exists only under certain conditions. In this case, the condition is one of *static equilibrium*: the end-point forces must balance the transverse load. Provided equilibrium may be established, the solution still remains *non-unique*, as it is possible to translate the wire perpendicularly to its axis without affecting the equilibrium. This is manifested by a singular stiffness matrix. An obvious computational treatment is to force the displacement at one node to be some known value, for instance zero. Adding this "superfluous boundary condition" makes the problem solvable uniquely. (We might wish to consider that one boundary condition specification was in fact missing because of the linear dependence between the slopes; so in this way we are really just filling a void.) Another possibility is to add an artificial spring-to-ground at one node.

4.4   Overspecified boundary conditions

The boundary condition application consisted so far of one and only one condition specification at each of the two endpoints (i.e. at *the boundary*). In this section we will attempt to apply *two boundary conditions at one end* and none at the other. Since this will turn out to be too much prescribed information at one point, we will call this the case of over-specified boundary conditions.

For instance, we could prescribe two pieces of information at $x = 0$

$$w(0) = \bar{w}_0 \; , \qquad \frac{\partial w}{\partial x}(0) = \bar{w}_0' \; , \qquad (4.9)$$

Integrating (4.2) once yields for the slope

$$\frac{\partial w}{\partial x}(x) = kx + C \; ,$$

which may be used in conjunction with the natural boundary condition to express the constant C as

$$C = \bar{w}_0' \; .$$

Integrating again yields

$$w(x) = k\frac{x^2}{2} + \bar{w}_0' x + w_0 \; ,$$

where all the constants are determined from the two boundary conditions at a single endpoint. Using this expression, we can calculate the deflection and slope at $x = L$. The interesting thing is that the slope is in general nonzero

$$\frac{\partial w}{\partial x}(L) = kL + \bar{w}_0' \neq 0 \; ,$$

which means that even though we did not assume this, there must have been an applied force F_L at the end where no boundary condition was specified. For the same reasons, it is not possible to prescribe either deflection or slope at $x = L$ while the boundary condition (4.9) is active. Any mismatch between the prescribed values and the calculated values would make the existence of the solution an impossibility (the solution must satisfy the balance equation and all boundary conditions).

The preceding discussion was based on an analytical integration of the balance equation, but is this way of prescribing boundary conditions compatible with the Galerkin technique? For the moment, we will go back all the way to the weighted residual equation (2.11), but we will keep both boundary terms (and drop the inertial terms – statics)

$$\eta_j(L)F_L - \eta_j(0)F_0 - \int_0^L \frac{\partial \eta_j}{\partial x} P \frac{\partial w}{\partial x} \, \mathrm{d}x + \int_0^L \eta_j q \, \mathrm{d}x = 0, \quad j = 1, \ldots, N \; .$$

$$(4.10)$$

The functions η_j and w need to be just sufficiently smooth to make the integrals exist. Now we introduce the boundary condition (4.9). As discussed below equation (2.10), the condition $w(x = 0) = \bar{w}_0$ could be used to eliminate the associated reaction, F_0, by setting $\eta_j(0) = 0$ for all j. The resulting weighted residual formulation is

$$\eta_j(L)F_L - \int_0^L \frac{\partial \eta_j}{\partial x} P \frac{\partial w}{\partial x}\, dx + \int_0^L \eta_j q\, dx = 0, \quad j = 1, \ldots, N \ . \quad (4.11)$$

where we must place one condition on w

$$w(x = 0) = \bar{w}_0, \quad w \in C^0 \ . \quad (4.12)$$

However, we realize that the prescribed force F_0 now no longer affects the solution! Another difficulty is that F_L is not known, and hence there is not enough equations to solve for $w_i, i = 2, \ldots, N$ and F_L. What is needed is an additional equation that would link together F_0 and F_L. The equation of global equilibrium is such an equation

$$F_L = F_0 - qL \ ,$$

and may be added to the equations resulting from (4.11). The global equations are then of a blocked nature. The Matlab script `woverspec` [1] generates a numerical solution for the definition of boundary conditions discussed above. The code snippet below illustrates the system matrix, `A`, in line 0034, which clearly displays how it is composed of distinct blocks

$$A = \begin{bmatrix} K & \begin{bmatrix} 0 \\ 0 \\ \vdots \\ 1 \end{bmatrix} \\ \begin{bmatrix} 0 \cdots 0 \end{bmatrix} & 1 \end{bmatrix} \ .$$

```
0028 K = start (dense_sysmat, get(w, 'neqns'));
0029 K = assemble (K, stiffness(feb, geom, w));
0030 % and now the special terms due to the boundary conditions
0032 z = zeros(N-1, 1); zt=z';
0033 z(end) = 1;
0034 A = [get(K,'mat'),z;zt,1];
```

---

[1] Folder: SOFEA/examples/taut_wire

The right-hand side of the system equations is correspondingly expanded by a single element at the bottom.

```
0036 fi = force_intensity(struct ('magn',inline(num2str(q))));
0037 F = start (sysvec, get(w, 'neqns'));
0038 F = assemble (F, body_loads(feb, geom, w, fi));
0039 F = assemble (F, nz_ebc_loads(feb, geom, w));
0040 % and now the special terms
0041 b=get(F,'vec');
0042 b = [b; -(F_0-q*L)];
```

Figure 4.1 illustrates the correct solution versus an incorrect solution obtained when the F_L term is simply dropped from (4.11): the natural boundary condition $F_L = 0$ is in effect. That can be verified visually as the slope of the incorrect solution at the right-hand end is apparently approaching zero.

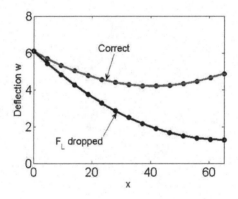

Fig. 4.1   The wire problem with overspecified boundary conditions.

Clearly, we can see that overspecification of boundary conditions does not fit the framework of the Galerkin method very well: specialized treatment is required. However, there is more bad news: the existence of the solution has to be demonstrated case-by-case, it does not follow automatically. The situation gets substantially worse in higher dimensional problems.

Chapter 5

Model of Heat Conduction

In this chapter we will develop a finite element model for heat conduction problems. The excellent textbook by Lienhard and Lienhard [Lienhard, Lienhard (2005)] has all the details one might require to supplement the treatment that follows.

5.1 Balance equation

In this section, our goal is to derive the balance equation that describes heat conduction in solids as a partial differential expression. It will be converted to a residual form, which will then be treated with the Galerkin method.

To begin, we pick a control volume, and we keep track of the heat energy within that volume. The control volume may be the whole structure, part of the structure, or just a very small chunk of material surrounding a given point in space (Fig. 5.1). The amount of heat energy in the control volume U is expressed as an integral of the volume density of heat energy, u

$$U = \int_V u \, dV \qquad (5.1)$$

The amount of heat energy within the control volume may change by

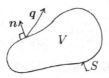

Fig. 5.1   The domain for the heat conduction problem.

outflow (inflow) of heat energy via the boundaries, and *heat generation* (or loss) within the volume. These quantities will be expressed in terms of rates. Therefore, the amount of energy flowing out of the control volume through its bounding surface S per unit time is

$$\int_S \boldsymbol{n} \cdot \boldsymbol{q} \, dS \, , \tag{5.2}$$

where $\boldsymbol{n}$ is the outer normal to the surface S, and $\boldsymbol{q}$ is the heat flux (amount of heat flowing through a unit area per unit time). The amount of energy generated within the control volume per unit time is

$$\int_V Q \, dV \, , \tag{5.3}$$

where Q is the rate of heat generation per unit volume; for example, heat is released or consumed by various deformation and chemical processes (as work of viscous stresses, reaction product of curing concrete or polymer resins, and so on).

Collecting the terms, we can write for the change of the heat energy within the control volume the rate equation

$$\frac{dU}{dt} = -\int_S \boldsymbol{n} \cdot \boldsymbol{q} \, dS + \int_V Q \, dV \, . \tag{5.4}$$

Finally, differentiating U with respect to time will be possible if we assume that $U = U(T)$, i.e. if U is a function of the absolute temperature T. Holding the control volume fixed in time, the time differentiation may be taken inside the integral over the volume

$$\frac{dU}{dt} = \frac{d}{dt} \int_V u \, dV = \int_V \frac{du}{dt} \, dV \, , \tag{5.5}$$

and with the application of the chain rule, the relationship (5.5) is expressed as

$$\frac{dU}{dt} = \int_V \frac{du}{dt} \, dV = \int_V \frac{du}{dT} \frac{\partial T}{\partial t} \, dV \, . \tag{5.6}$$

The quantity $c_V = du/dT$ is a characteristic property of a solid material (called specific heat at constant volume). It is typically dependent on temperature, but we will assume that it is a constant; otherwise it leads to nonlinear models.

Substituting, we write

$$\int_V c_V \frac{\partial T}{\partial t}\, dV = -\int_S \boldsymbol{n} \cdot \boldsymbol{q}\, dS + \int_V Q\, dV\ . \tag{5.7}$$

This equation consists of volume integrals and a surface integral. If all the integrals were volume integrals, over the same volume of course, we could proclaim that the integral statement (sometimes called a global balance equation) would hold provided the integrands satisfied a so-called local balance equation (recall that to get the local balance equation is our goal). For instance, from the integral statement

$$\int_V \alpha \frac{\partial M}{\partial t}\, dV = \int_V \mu\, dV\ , \tag{5.8}$$

where α, M, and μ are some functions, one could conclude that

$$\alpha \frac{\partial M}{\partial t} = \mu\ , \tag{5.9}$$

which is a local version of (5.8). An argument along these lines could for instance invoke the assumption that the volume V was arbitrary, and that it could be shrunk around a given point, which in the limit would allow the volume to be canceled on both sides of the equation.

To execute this program for Eq. (5.7), we have to convert the surface integral to a volume integral. We have the needed tool in the celebrated **divergence theorem**

$$\int_V \operatorname{div}\boldsymbol{q}\, dV = \int_S \boldsymbol{n} \cdot \boldsymbol{q}\, dS\ , \tag{5.10}$$

where the divergence of the flux vector is defined in Cartesian coordinates as

$$\operatorname{div}\boldsymbol{q} = \frac{\partial q_x}{\partial x} + \frac{\partial q_y}{\partial y} + \frac{\partial q_z}{\partial z}\ .$$

Consequently, Eq. (5.7) may be rewritten

$$\int_V c_V \frac{\partial T}{\partial t}\, dV = -\int_V \operatorname{div}\boldsymbol{q}\, dV + \int_V Q\, dV\ , \tag{5.11}$$

and grouping the terms as

$$\int_V \left[c_V \frac{\partial T}{\partial t} + \operatorname{div}\boldsymbol{q} - Q \right] dV = 0\ . \tag{5.12}$$

we may conclude that the inside of the bracket has to vanish since the volume could be entirely arbitrary. Therefore, we arrive at the **local balance equation**

$$c_V \frac{\partial T}{\partial t} + \text{div} \boldsymbol{q} - Q = 0 \ . \tag{5.13}$$

5.2   Constitutive equation

Equation (5.13) contains too many variables: both temperature and heat flux. Since it is a scalar equation, the logical step is to express the heat flux in terms of temperature. That is the content of the Fourier model: heat flows opposite to the gradient of the temperature (downhill). In matrix form

$$\boldsymbol{q} = -\boldsymbol{\kappa}(\text{grad}T)^T \ . \tag{5.14}$$

The matrix $\boldsymbol{\kappa}$ is the conductivity matrix of the material. The most common forms of $\boldsymbol{\kappa}$ are

$$\boldsymbol{\kappa} = \kappa \mathbf{1} \tag{5.15}$$

for the so-called thermally isotropic material, and

$$\boldsymbol{\kappa} = \begin{pmatrix} \kappa_x & 0 & 0 \\ 0 & \kappa_y & 0 \\ 0 & 0 & \kappa_z \end{pmatrix} , \tag{5.16}$$

for materials that have three orthogonal directions of different thermal conductivities (orthotropic material); κ is the isotropic thermal conductivity coefficient, $\mathbf{1}$ is the identity matrix, and κ_x, κ_y, and κ_z are the orthotropic thermal conductivities. To explain the orthotropic conductivity model we note that some materials have preferred directions in which heat would like to flow, for instance along the fibers in a composite. Visually, we can imagine a corrugated steel roof, with the channels running not directly downhill, but tilted away from the slope – the water would run preferentially in the channels, but generally downhill.

The funny looking transpose of the temperature gradient follows from the definition: the gradient of the scalar is a row matrix

$$\text{grad}T = \left[\frac{\partial T}{\partial x}, \frac{\partial T}{\partial y}, \frac{\partial T}{\partial z} \right] \ . \tag{5.17}$$

With the constitutive equation, the balance equation (5.13) is now expressed purely in terms of the absolute temperature,

$$c_V \frac{\partial T}{\partial t} - \text{div} \left[\boldsymbol{\kappa} (\text{grad} T)^T \right] - Q = 0 \; . \tag{5.18}$$

5.3 Boundary conditions

From now on, V is going to be the volume of the whole solid domain. The most important fact about the boundary conditions is that we need to have a boundary condition *at each point* of the surface S. As we may suspect by now, the model is all about temperature. Correspondingly, the boundary conditions are an expression of our *a priori* knowledge of the temperature distribution in the solid or on its surface.

The simplest boundary condition results if we know the surface temperature along one part of S at all times. This part of the surface will be called S_1 (see Fig. 5.2). Therefore,

$$T(\boldsymbol{x}, t) - \overline{T}(\boldsymbol{x}, t) = 0, \quad \boldsymbol{x} \text{ on } S_1 \; . \tag{5.19}$$

This type of condition is known as the primary, or **essential**, boundary condition.

The heat flux entering or leaving the solid may also be known (measured by a heat flux gauge, for instance). Generally, we do not know the heat flux along the surface, only the normal component, which is available from the normal to the surface and the heat flux as $q_n = \boldsymbol{n} \cdot \boldsymbol{q}$. Therefore, along the S_2 part of the surface the normal component of the heat flux may be prescribed

$$\boldsymbol{n} \cdot \boldsymbol{q} - \overline{q}_n = 0, \quad \boldsymbol{x} \text{ on } S_2 \; . \tag{5.20}$$

All quantities are given at a particular point on the boundary as functions of time, similarly to the first boundary condition. This type of condition is known as the **natural** (or flux) boundary condition.

As the last example of a boundary condition, we will mention heat transfer driven by a temperature difference across a surface. The normal component of the heat flux is given as

$$\boldsymbol{n} \cdot \boldsymbol{q} - h(T - T_a) = 0, \quad \boldsymbol{x} \text{ on } S_3 \; , \tag{5.21}$$

where T_a is the known temperature of the surrounding medium (ambient temperature), and h is the heat transfer coefficient.

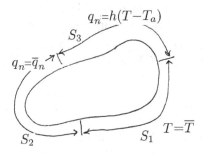

Fig. 5.2   The partitioning of the boundary surface used in the description of the boundary conditions.

5.3.1   *On the sufficiency of boundary conditions.*

As pointed out earlier in this section, one boundary condition is needed at each point on the boundary. The precise mathematical statement of the necessity of having one boundary condition in place is somewhat involved, but we can build on intuition fairly easily.

Would it be possible to specify one boundary condition at only a subset of the complete boundary, leaving the behavior of the solution along a part or parts of the boundary unspecified? As a thought experiment, we consider a square domain, shown in Fig. 5.3, with no source of heat generation, and zero temperature prescribed on the S_1 subset of the boundary. On the $\widehat{S}$ part of the boundary we assume nothing is known about the temperature distribution. Is it possible that the temperature field is completely determined by these boundary conditions?

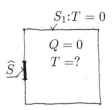

Fig. 5.3   The square domain with partially undefined boundary condition.

If it was true, the variation of temperature along $\widehat{S}$ wouldn't affect the solution in the domain. However, if zero temperature was prescribed all around the circumference of the square, the solution to this problem would be zero temperature everywhere. Consequently, also the normal component of the flux (in fact all components of the flux) would vanish everywhere. Evidently, if the temperature along $\widehat{S}$ was nonzero, it would require transitioning to zero temperature on S_1 (and elsewhere within the domain), hence the solution within the domain *would* depend on the temperature along $\widehat{S}$; alternatively, if there was nonzero heat flux along $\widehat{S}$, the temperature distribution within the square domain would be affected. This is illustrated in Fig. 5.4: varying the heat flux along $\widehat{S}$ (here shown for two different uniform distributions, positive and negative) changes the distribution of temperature. Therefore, we must conclude that *prescribing one boundary condition along the entire boundary* of the domain is a *necessary condition* for making the solution unique.

Is not a sufficient condition, however. In the so-called Neumann problem only the heat flux is being prescribed along the entire boundary. This is equivalent to the pure-traction problem of Section 4.3. The solution is not unique, because any temperature distribution of the form $T(x, y) + \widetilde{T}$, where $T(x, y)$ satisfies the balance equation and the natural boundary conditions, and $\widetilde{T}$ is a constant, is also a solution (the constant term disappears with differentiation). Typically, the Neumann boundary conditions are supplemented with temperature being prescribed at one point to remove the constant $\widetilde{T}$ from consideration.

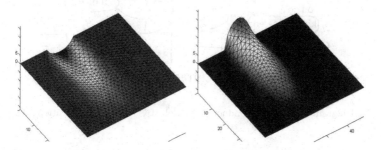

Fig. 5.4 The square with two different distributions of heat flux along $\widehat{S}$.

5.4 Initial condition

The primary variable in our problem is the temperature, T, and it is present in the balance equation (5.18) with the first order time derivative. Therefore, we will need one initial condition,

$$T(\boldsymbol{x}, 0) = \overline{T}_0(\boldsymbol{x}) \quad \boldsymbol{x} \text{ in } V \ . \tag{5.22}$$

The initial condition must match any boundary condition on S_1 at time $t = 0$:

$$\overline{T}_0(\boldsymbol{x}) = \overline{T}(\boldsymbol{x}, 0), \quad \boldsymbol{x} \text{ on } S_1 \ . \tag{5.23}$$

5.5   Summary of the PDE model of heat conduction

Figure 5.5 gives a pictorial overview of the terminology and the various equations of the model of heat conduction (for the curious: it is a simplified Tonti diagram). One of the main points of a picture is to save a thousand words, so I let it do its magic.

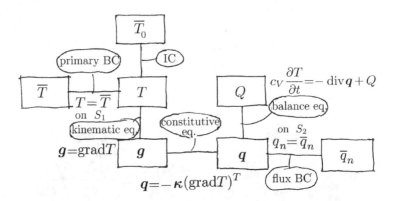

Fig. 5.5   Diagram of the heat conduction model. BC: boundary condition; IC: initial condition.

Exercises

(1) List the types of the boundary conditions for the heat conduction problem that you can find in the open literature – textbooks, papers, and on the Web. Give an equation to describe the boundary condition, and discuss a physical situation in which the boundary condition may be appropriate. Also, indicate whether the boundary condition will lead to a linear model or to a nonlinear model.

Chapter 6

Galerkin Method for the Model of Heat Conduction

6.1 Weighted residual formulation

The balance equation (5.18) yields the balance residual as

$$r_B = c_V \frac{\partial T}{\partial t} - \operatorname{div}\left[\kappa(\operatorname{grad}T)^T\right] - Q . \tag{6.1}$$

As explained in Chapter 2, the first step is to satisfy the essential boundary condition by restricting possible trial functions to only those that conform to the essential boundary conditions a priori

$$T(\boldsymbol{x}, t) - \overline{T}(\boldsymbol{x}, t) = 0, \quad \boldsymbol{x} \text{ on } S_1 , \tag{6.2}$$

and to write the weighted residual equation for the local balance residual

$$\int_V \eta(\boldsymbol{x}) r_B(\boldsymbol{x}, t) \, \mathrm{d}V = 0 . \tag{6.3}$$

The first and the third term are kept without change, but the second term reminds us of a similar term in Eq. (2.3): the test function η multiplies an expression that contains the second derivatives of temperature (the $\operatorname{div}\left[\kappa(\operatorname{grad}T)^T\right]$ term). Balancing the order of differentiation by shifting one derivative from the temperature to the test function η will be beneficial: similarly to Section 2.5, we will be able to satisfy the natural boundary conditions without having to include them as a residual (naturally!). As before, the price to pay is the need to place some restrictions on the test function.

Integration by parts was used in Section 2.5, and just a little bit more general tool will work here too. For the moment, it will be convenient to work with the expression

$$-\eta \ \mathrm{div} \left[\boldsymbol{\kappa}(\mathrm{grad}T)^{T}\right] = \eta \ \mathrm{div}\boldsymbol{q} \ ,$$

that is, we revert to the flux variable instead of $-\boldsymbol{\kappa}(\mathrm{grad}T)^{T}$.

The integration by parts in the case of a multidimensional integral is generalized in the divergence theorem (5.10). We may anticipate that $\eta \ \mathrm{div}\boldsymbol{q}$ is the result of the chain rule applied to the vector $\eta \ \boldsymbol{q}$. That is indeed the case,

$$\mathrm{div} \left(\eta \ \boldsymbol{q}\right) = \eta \ \mathrm{div}\boldsymbol{q} + (\mathrm{grad}\eta) \ \boldsymbol{q} \ , \tag{6.4}$$

which is easily verified in components. Therefore, we may start by inspecting the integral

$$\int_{V} \eta \ \mathrm{div}\boldsymbol{q} \ \mathrm{d}V$$

where we substitute from (6.4)

$$\int_{V} \eta \ \mathrm{div}\boldsymbol{q} \ \mathrm{d}V = \int_{V} \mathrm{div} \left(\eta \ \boldsymbol{q}\right) \ \mathrm{d}V - \int_{V} (\mathrm{grad}\eta) \ \boldsymbol{q} \ \mathrm{d}V \ . \tag{6.5}$$

The divergence theorem may be applied to the first integral on the right

$$\int_{V} \eta \ \mathrm{div}\boldsymbol{q} \ \mathrm{d}V = \int_{S} \eta \ \boldsymbol{q} \cdot \boldsymbol{n} \ \mathrm{d}S - \int_{V} (\mathrm{grad}\eta) \ \boldsymbol{q} \ \mathrm{d}V \ . \tag{6.6}$$

Since $\boldsymbol{q} \cdot \boldsymbol{n}$ is known on some parts of the boundary, but unknown on the others – see Eqs. (5.20) and (5.21), we will split the surface integral into one for each sub-surface,

$$\int_{V} \eta \ \mathrm{div}\boldsymbol{q} \ \mathrm{d}V = \int_{S_1} \eta \ \boldsymbol{q} \cdot \boldsymbol{n} \ \mathrm{d}S + \int_{S_2} \eta \ \bar{q}_n \ \mathrm{d}S + \int_{S_3} \eta \ h(T - T_a) \ \mathrm{d}S$$
$$- \int_{V} (\mathrm{grad}\eta) \ \boldsymbol{q} \ \mathrm{d}V \ . \tag{6.7}$$

We see that the situation is analogous to the one discussed below Eq. (2.10): The integral over the part of the surface S_1 is troublesome, because $\boldsymbol{q} \cdot \boldsymbol{n}$ is unknown there. However, we have the option of making η vanish along S_1. In this way, we obtain

$$\int_{V} \eta \ \mathrm{div}\boldsymbol{q} \ \mathrm{d}V = \int_{S_2} \eta \ \bar{q}_n \ \mathrm{d}S + \int_{S_3} \eta \ h(T - T_a) \ \mathrm{d}S - \int_{V} (\mathrm{grad}\eta) \ \boldsymbol{q} \ \mathrm{d}V \ , \tag{6.8}$$

where $\eta(\boldsymbol{x}) = 0$ for $\boldsymbol{x} \in S_1$. Expanding the weighted residual equation (6.3) yields

$$\int_V \eta \, r_B \, dV =$$
$$\int_V \eta c_V \frac{\partial T}{\partial t} \, dV + \int_V (\mathrm{grad}\eta) \, \boldsymbol{\kappa}(\mathrm{grad}T)^T \, dV - \int_V \eta Q \, dV$$
$$+ \int_{S_2} \eta \, \bar{q}_n \, dS + \int_{S_3} \eta \, h(T - T_a) \, dS = 0, \quad \eta(\boldsymbol{x}) = 0 \text{ for } \boldsymbol{x} \in S_1 \, .$$

$$(6.9)$$

In this equation we have our result: a single weighted residual statement with balanced derivatives.

6.2 Reducing the model dimension

In this section we show how the originally three-dimensional model can be reduced to just two active dimensions. (The reduced model will still describe the heat conduction through a *three-dimensional domain*; the function describing the temperature distribution will depend only on *two* spatial variables though.)

For some physical situations we can make the observation that the temperature does not vary significantly along one coordinate direction, say along the z direction. Figure 6.1 shows a disk of thickness Δz. It is a slice of a structure of an unchanging cross-section which is very along in the z direction compared to the transverse dimensions. If we can neglect what is happening near the end sections, and if the component of the temperature gradient along the z direction is negligible, $\partial T/\partial z \approx 0$, a necessary condition for the formulation of a simplified model is met. However, it is not a sufficient condition as it does not necessarily mean that the z component

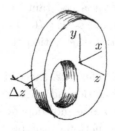

Fig. 6.1   Slice of a long cylindrical structure

of the heat flux is also zero: the partial derivatives $\partial T/\partial x$, and $\partial T/\partial y$ multiply the first two columns in row three of (5.14) to yield

$$q_z = \kappa_{zx}\partial T/\partial x + \kappa_{zy}\partial T/\partial y \ .$$

However, for the two classes of materials (5.15) and (5.16) the two coefficients κ_{zx} and κ_{zy} are identically zero, which means that if the temperature gradient $\partial T/\partial z$ is zero, the heat flux in that direction also vanishes.

Going back to Fig. 6.1: the heat flux through the cross sections is zero, and the temperature through the thickness of the disk is uniform (i.e. the temperature does not vary with z). The surface of the three-dimensional solid consists of the two cross sections, and of the cylindrical surfaces, the inner and the outer. The two cylindrical surfaces may be associated with boundary condition of any type. The two cross sections are associated with the boundary condition of zero heat flux, $\overline{q}_n = 0$ (type S_2, Eq. (5.20))

$$\boldsymbol{n} \cdot \boldsymbol{q} = \pm q_z = 0, \quad \text{on the cross sections .} \tag{6.10}$$

Since the temperature does not vary with z, the integrals (6.8) may be simplified by pre-integrating in the thickness direction, $\mathrm{d}V = \Delta z \, \mathrm{d}S$. The volume integrals are then evaluated over the cross-sectional area, S_c, (see Fig. 6.2); provided $\overline{q}_n$ and h are independent of z, the surface integrals are computed as integrals over the contour of the cross-section, C_c.

$$\int_{S_c} \eta c_V \frac{\partial T}{\partial t} \, \Delta z \, \mathrm{d}S + \int_{S_c} (\mathrm{grad}\eta) \, \boldsymbol{\kappa}(\mathrm{grad}T)^T \, \Delta z \, \mathrm{d}S - \int_{S_c} \eta Q \, \Delta z \, \mathrm{d}S$$
$$+ \int_{C_{c,2}} \eta \, \overline{q}_n \, \Delta z \, \mathrm{d}C + \int_{C_{c,3}} \eta \, h(T - T_a) \, \Delta z \, \mathrm{d}C = 0,$$
$$\eta(\boldsymbol{x}) = 0 \text{ for } \boldsymbol{x} \in C_{c,1} \ .$$

$$\tag{6.11}$$

Note that the thickness Δz is a constant and could cancel without any effect on the solution. Nevertheless, Eq. (6.11) still applies to a fully three-dimensional body. To maintain this notion throughout the book, we shall not cancel the thickness.

Note that (6.11) does not refer to z, except in the term $\partial./\partial z$. We know that the temperature does not depend on z, and concerning the gradient of η: we simply assume that η does not depend on z: $\eta = \eta(x,y)$. The last assumption completes the reduction of the problem to two dimensions: all the functions depend on x and y only.

Fig. 6.2   Diagram of the heat conduction model.

6.3   Test and trial functions: basis functions on triangulations

It is time to talk about the test and trial function. They are both functions of x and y only, $\eta = \eta(x, y)$ and $T = T(x, y, t)$ (and for the trial function, time). The only difference between them is the value they assume on one part of the boundary (which is a part of the cross-section contour, for our two-dimensional disk) where the temperature is being prescribed, $C_{c,1}$:

$$T(\boldsymbol{x}, t) = \overline{T}(\boldsymbol{x}, t), \quad \eta(\boldsymbol{x}) = 0, \quad \boldsymbol{x} \text{ on } C_{c,1}.$$

Let us consider first the test function. It needs to be defined as a function of x and y over arbitrarily shaped domains. The concept of piecewise linear functions defined over tilings of arbitrary domains into triangles is quite ancient (at least in terms of the development of computational mechanics). The so-called "linear triangle" made its first appearance in a lecture by Courant in 1943, applied to Poisson's equation, which is a time-independent version of the heat conduction equation of this chapter. It was then picked up as a structural element in aerospace engineering to model Delta wing skin panels, as described in the 1956 paper by Turner, Clough, Martin and Topp. Clough then applied the triangle to problems in civil engineering, and he also coined the terminology "finite element".

The domain of the disk with a hole (shown in Fig. 6.2) is approximated as a collection of triangles (in other words, it is *tiled* with triangles, or *triangulated*), see Fig. 6.3. The mesh consisting of triangles is typically called **triangulation**, even though sometimes *any* mesh is called that. The vertices of the triangulation are called **nodes** (compare with Section 2.7), while the line segments connecting the nodes are called **edges**. Evidently, the triangles are the finite elements.

Interpolation on the triangle mesh will be treated as a linear combination of "tent" functions. Each individual tent is formed by grabbing one

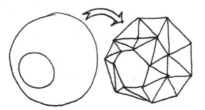

Fig. 6.3   Mesh of the disk domain.

of the nodes (say J) and raising it out of the plane of the triangulation (traditionally to a unit height). The tent canvas is stretched over the edges that connect at the node J, and are clamped down by the ring of the edges that surround node J. The cartoon of one particular basis function tent is shown in Fig. 6.4. For those who do not like tents (perhaps it rained a lot during the summer camp), the term *hat function* may be preferable.

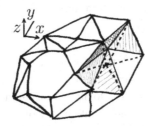

Fig. 6.4   Visual representation of one basis function on the mesh of the disk.

All the triangles that are connected in the node J **support** the function N_J, which is another way of saying that the function N_J is nonzero in these triangles; it is defined to be zero everywhere else. (If we are inside the "tent", we are standing on the support of the function.) Mathematically, the support of the basis function N_J is

$$\text{supp} N_J = \{\boldsymbol{x} : N_J(\boldsymbol{x}) \neq 0\} .$$

Since the set suppN_J is a finite piece of the (typically finite) computational domain, it is also called a **compact support**. The compact supports of the finite element basis functions make the finite element matrices sparse, and hence are crucial for the efficiency of these methods.

It remains to write down the equations that define the function N_J at any point within its support. That means writing an expression for each triangle separately. As discussed in Section 2.9, the alternative viewpoint would rather express all the nonzero pieces of all the basis functions over a single triangle (element). Referring to Fig. 6.4, there are only three such functions: the three basis functions associated with the nodes at the corners of the element; all the other basis functions in the mesh are identically zero over this element. Thus, our task is to write down the expressions for the three basis functions over the domain of a single triangle.

6.4 Basis functions on the standard triangle

Each of the three basis functions is zero along one edge of the triangle: again, refer to Fig. 6.4. The task is accomplished most readily when the triangle is in a special position with respect to the coordinates: the **standard triangle**; see Fig. 6.5. The basis functions associated with nodes ② and ③ are simply

$$N_2(\xi, \eta) = \xi \, , \tag{6.12}$$

and

$$N_3(\xi, \eta) = \eta \, . \tag{6.13}$$

As is easily verified, N_2 is zero along the edge ①③, and assumes value $+1$ at node ②; analogous properties hold for N_3. If N_1 should be equal to $+1$ at the origin, it must be written as

$$N_1(\xi, \eta) = 1 - \xi - \eta \, . \tag{6.14}$$

Clearly, N_1 vanishes at the edge opposite node ①. Thus, we see that the three functions we just formulated satisfy the Kronecker delta property, equation (2.21). As in Section 2.7, this means the degree of freedom at each node of the triangle is the value of the interpolated function at the node.

Also, we have the following property of the **partition of unity**

$$\sum_{k=1}^{3} N_k(\xi, \eta) = 1 \, , \tag{6.15}$$

which should be interpreted in this sense: the basis functions "partition" $+1$ at any point within the triangle, and we will make use of this property later to show which functions will be reproduced exactly when interpolated over an element.

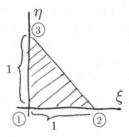

Fig. 6.5   Standard triangle.

As the three functions satisfy the Kronecker delta property (2.21), the degree of freedom at each node of the triangle is the value of the interpolated function at the node, $T_i = T(\boldsymbol{x}_i)$. Therefore, we may make the observation that data that sit at the corners of the triangle are **naturally interpolated**. One particularly useful quantity that one can interpolate on the standard triangle are the Cartesian coordinates of the corners in the physical space,

$$\boldsymbol{x} = \sum_{i=1}^{3} N_i(\xi, \eta) \boldsymbol{x}_i \;, \tag{6.16}$$

where the result of the interpolation is a point in the Cartesian coordinates

$$\boldsymbol{x} = \begin{bmatrix} x \\ y \end{bmatrix} \;,$$

and

$$\boldsymbol{x}_i = \begin{bmatrix} x_i \\ y_i \end{bmatrix} \quad i = 1, 2, 3 \;,$$

are the coordinates of the three points that are being interpolated. Equation (6.16) is a mapping from the pair ξ, η to the point x, y. Substituting for the basis functions, it may be written explicitly as

$$\begin{bmatrix} x \\ y \end{bmatrix} = \begin{bmatrix} (x_2 - x_1) & (x_3 - x_1) \\ (y_2 - y_1) & (y_3 - y_1) \end{bmatrix} \begin{bmatrix} \xi \\ \eta \end{bmatrix} + \begin{bmatrix} x_1 \\ y_1 \end{bmatrix} \;. \tag{6.17}$$

This matrix equation is accompanied by the picture in Fig. 6.6. The two vectors, $\boldsymbol{v}$ and $\boldsymbol{w}$, are the two columns of the square matrix in (6.17):

$$\boldsymbol{v} = \begin{bmatrix} (x_2 - x_1) \\ (y_2 - y_1) \end{bmatrix} \;,$$

and

$$w = \begin{bmatrix} (x_3 - x_1) \\ (y_3 - y_1) \end{bmatrix}.$$

If both ξ and η vary between zero and one, equation (6.17) adds the two vectors, ξv and ηw to the vector $[x_1, y_1]^T$, and the result then covers the entire parallelogram; on the other hand, if ξ and η are confined to the interior of the standard triangle, Eq. (6.17) produces points to cover the area of the hatched triangle. To summarize, Eq. (6.17) is a **map** from the standard triangle to a triangle in the Cartesian coordinates with corners in given locations.

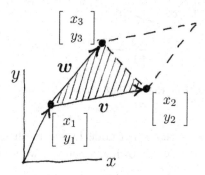

Fig. 6.6   Interpolating Cartesian coordinates on the standard triangle.

Inverting (6.17) to express ξ and η, which could then be substituted into (6.12) – (6.14) to produce basis functions in terms of x and y, may look appealing but should be resisted. The reason is that numerical quadrature is available on the standard triangle, but is much harder on general triangles. This will become especially clear with quadratic elements later in the book.

However, since Eq. (6.17) is an invertible map from the standard triangle to a triangle in the Cartesian coordinates (invertibility follows if the triangle does not have its corners in a single straight-line: why?), we do get an algorithm for *evaluating basis functions on a general triangle*. Given a point $\bar{x}, \bar{y}$ in the Cartesian coordinates, and within the bounds of a triangle, we can use the inverse of the map (6.17) to obtain point $\bar{\xi}, \bar{\eta}$ in the standard triangle (path 1 in Fig. 6.7). Therefore, we may then evaluate $N_i(\bar{\xi}, \bar{\eta})$, which is the value $N_i(\bar{x}, \bar{y})$ (path 2 in Fig. 6.7). That may seem awkward, but normally we would want to evaluate the basis functions in order to perform numerical quadrature, that is at a particular point (quadrature point) within the standard triangle. In that case, $\bar{\xi}, \bar{\eta}$ would be *known* (and $\bar{x}, \bar{y}$ would be unknown), and calculation of the function value is easy. Evaluation of the *derivatives* of the basis functions is a little bit more complex, and will be therefore discussed separately in the section on numerical quadrature.

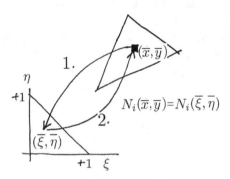

Fig. 6.7   Using the map from the standard triangle to evaluate basis functions over a general triangle.

We understand now that each node in the mesh is associated with a single basis function. In the following, whenever we write

$$N_i = N_i(x, y) \ ,$$

it has to be understood that within each triangle in the mesh, the coordinates of the point are given as $x = x(\xi, \eta)$, $y = y(\xi, \eta)$, where ξ and η are coordinates in the standard triangle.

6.5   Discretizing the weighted residual equation

The trial function will be expressed using the basis functions as (compare with Section 2.7)

$$T(x, y, t) = \sum_{i=1}^{N} N_i(x, y) T_i(t) \ ,$$

where the sum ranges over all the basis functions (i.e. over all the nodes in the mesh). Included are also basis functions associated with the nodes on the boundary where the temperature is being prescribed, $C_{c,1}$. On the contrary, these nodes do not contribute basis functions to the set of the test functions (these are expected to vanish along $C_{c,1}$). Therefore, we will choose

$$\eta(x,y) = N_i(x,y), \quad i \text{ excluded when node } i \in C_{c,1} \,.$$

The nodes whose basis functions are not part of the linear combination for the test function are shown as empty circles in Fig. 6.8.

To simplify, we shall adopt the following notation:

$$\eta(x,y) = N_j(x,y), \quad \forall \text{ free } j,$$

where "free j" ranges over the nodes where the temperature is not being prescribed; and

$$T(x,y,t) = \sum_{\text{all } i} N_i(x,y)T_i(t)\,,$$

where "all i" ranges over all the nodes, including those where the temperature is being prescribed. In addition, because the basis on the standard triangle satisfies the Kronecker delta property (2.21), the values of the degrees of freedom $T_i(t)$ at the nodes "prescribed i" (the nodes with the empty circles in Fig. 6.8) are simply the values of the interpolated prescribed temperature at the nodes, $T_i(t) = \overline{T}(x_i, y_i, t)$.

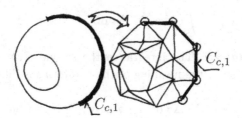

Fig. 6.8 Interpolating Cartesian coordinates on the standard triangle.

Remark: Apropos curved boundaries: Figure 6.8 clearly shows that with straight edges we are only approximating any boundaries that are curved. Some error is involved, but fortunately we are able to control this error by reducing the length of the edges.

The finite element expansions for the trial and test functions are now substituted into the weighted residual integral (6.11). For clarity, the substitution will be shown term-by-term (henceforth we will omit the arguments):

$$\int_{S_c} \eta c_V \frac{\partial T}{\partial t} \, \Delta z \, \mathrm{d}S = \int_{S_c} N_j c_V \sum_{\text{all } i} N_i \frac{\partial T_i}{\partial t} \, \Delta z \, \mathrm{d}S \ , \quad \forall \text{ free } j \ , \qquad (6.18)$$

which simplifies to

$$\sum_{\text{all } i} \left[\int_{S_c} N_j c_V N_i \, \Delta z \, \mathrm{d}S \right] \frac{\partial T_i}{\partial t} \ , \quad \forall \text{ free } j \ . \qquad (6.19)$$

The term in the bracket mixes together i and j from two different sets, and some of the degrees of freedom $\partial T_i / \partial t$ are known. Therefore, separating the known and unknown quantities may be a good idea:

$$\sum_{\text{all } i} \left[\int_{S_c} N_j c_V N_i \, \Delta z \, \mathrm{d}S \right] \frac{\partial T_i}{\partial t} =$$
$$\sum_{\text{free } i} \left[\int_{S_c} N_j c_V N_i \, \Delta z \, \mathrm{d}S \right] \frac{\partial T_i}{\partial t} + \qquad (6.20)$$
$$\sum_{\text{prescribed } i} \left[\int_{S_c} N_j c_V N_i \, \Delta z \, \mathrm{d}S \right] \frac{\partial \overline{T}_i(t)}{\partial t} \ , \quad \forall \text{ free } j \ .$$

The first integral on the right-hand side of (6.20) suggests defining a matrix

$$C_{ji} = \int_{S_c} N_j c_V N_i \, \Delta z \, \mathrm{d}S \ , \quad \forall \text{ free } j, i, \qquad (6.21)$$

the **capacity matrix.** The integral in the second term will be given a different symbol, since the meaning of the two terms is different

$$\overline{C}_{ji} = \int_{S_c} N_j c_V N_i \, \Delta z \, \mathrm{d}S \ , \quad \forall \text{ free } j, \ \forall \text{ prescribed } i. \qquad (6.22)$$

Next, the second term in (6.11):

$$\int_{S_c} (\mathrm{grad}\eta) \, \kappa (\mathrm{grad} T)^T \, \Delta z \, \mathrm{d}S = \int_{S_c} (\mathrm{grad} N_j) \, \kappa (\mathrm{grad} \sum_{\text{all } i} N_i T_i)^T \, \Delta z \, \mathrm{d}S =$$
$$\sum_{\text{free } i} \left[\int_{S_c} (\mathrm{grad} N_j) \, \kappa (\mathrm{grad} N_i)^T \, \Delta z \, \mathrm{d}S \right] T_i +$$
$$\sum_{\text{prescribed } i} \left[\int_{S_c} (\mathrm{grad} N_j) \, \kappa (\mathrm{grad} N_i)^T \, \Delta z \, \mathrm{d}S \right] \overline{T}_i \quad \forall \text{ free } j,$$

$$(6.23)$$

and the **conductivity matrix** may be defined as

$$K_{ji} = \int_{S_c} (\text{grad}N_j) \, \boldsymbol{\kappa}(\text{grad}N_i)^T \, \Delta z \, dS \,, \quad \forall \text{ free } j, i, \tag{6.24}$$

and the elements to go with the load-like term

$$\overline{K}_{ji} = \int_{S_c} (\text{grad}N_j) \, \boldsymbol{\kappa}(\text{grad}N_i)^T \, \Delta z \, dS \,, \quad \forall \text{ free } j, \, \forall \text{ prescribed } i. \tag{6.25}$$

Next, the load term corresponding to the internal heat generation:

$$L_{Q,j} = \int_{S_c} N_j Q \, \Delta z \, dS \,, \quad \forall \text{ free } j. \tag{6.26}$$

Finally, the terms corresponding to natural boundary conditions. On the $C_{c,2}$ part of the boundary, only a load term results.

$$L_{q2,j} = -\int_{C_{c,2}} N_j \, \overline{q}_n \, \Delta z \, dC \,. \tag{6.27}$$

On the other hand, on the $C_{c,3}$ part of the boundary, where the heat flux is proportional to the difference between the ambient temperature and the surface temperature, we get a load term

$$L_{q3,j} = \int_{C_{c,3}} N_j \, hT_a \, \Delta z \, dC \,, \quad \forall \text{ free } j, \tag{6.28}$$

and a **surface heat transfer matrix**:

$$H_{ji} = \int_{C_{c,3}} N_j \, hN_i \, \Delta z \, dC \,, \quad \forall \text{ free } j, i, \tag{6.29}$$

and

$$\overline{H}_{ji} = \int_{C_{c,3}} N_j \, hN_i \, \Delta z \, dC \,, \quad \forall \text{ free } j, \, \forall \text{ prescribed } i \,. \tag{6.30}$$

Also, we shall be writing

$$L_{\overline{H},j} = -\sum_{\text{prescribed } i} \overline{H}_{ji} \overline{T}_i \,, \quad \forall \text{ free } j, \tag{6.31}$$

$$L_{\overline{K},j} = -\sum_{\text{prescribed } i} \overline{K}_{ji} \overline{T}_i \,, \quad \forall \text{ free } j, \tag{6.32}$$

$$L_{\overline{C},j} = -\sum_{\text{prescribed } i} \overline{C}_{ji} \frac{\partial \overline{T}_i(t)}{\partial t} \ , \quad \forall \text{ free } j, \tag{6.33}$$

for the loads terms produced by nonzero essential boundary conditions.

To summarize, using the definitions of the various matrices and load terms, the system of ordinary differential equations that results from the finite element discretization in space reads

$$\sum_{\text{free } i} C_{ji} \frac{\partial T_i}{\partial t} + \sum_{\text{free } i} K_{ji} T_i + \sum_{\text{free } i} H_{ji} T_i$$

$$-L_{\overline{C},j} - L_{\overline{K},j} - L_{\overline{H},j} - L_{Q,j} - L_{q2,j} - L_{q3,j} = 0 \qquad \forall \text{ free } j. \tag{6.34}$$

6.6 Derivatives of the basis functions; Jacobian

The results of this section are much more general than may be expected. While the formulas for the derivatives of basis functions are derived for the linear triangles, the same formulas (and implementation) is used for all the so-called *isoparametric* elements in the SOFEA toolbox.

To evaluate the conductivity matrix, we need to be able to calculate the derivatives of the basis functions with respect to x and y. Equations (6.12–6.14) define the functions over the standard triangle in terms of ξ and η. Therefore, to express $\partial N_i / \partial x$ we use the chain rule

$$\frac{\partial N_i}{\partial x} = \frac{\partial N_i}{\partial \xi} \frac{\partial \xi}{\partial x} + \frac{\partial N_i}{\partial \eta} \frac{\partial \eta}{\partial x} \ ,$$

$$\frac{\partial N_i}{\partial y} = \frac{\partial N_i}{\partial \xi} \frac{\partial \xi}{\partial y} + \frac{\partial N_i}{\partial \eta} \frac{\partial \eta}{\partial y} \ .$$

For the purpose of this discussion, the function that is being differentiated does not really matter. We will replace it with a $\heartsuit$, while we arrange the above equation into a matrix expression

$$\left[\frac{\partial \heartsuit}{\partial x}, \ \frac{\partial \heartsuit}{\partial y} \right] = \left[\frac{\partial \heartsuit}{\partial \xi}, \ \frac{\partial \heartsuit}{\partial \eta} \right] \begin{bmatrix} \dfrac{\partial \xi}{\partial x} & \dfrac{\partial \xi}{\partial y} \\[2ex] \dfrac{\partial \eta}{\partial x} & \dfrac{\partial \eta}{\partial y} \end{bmatrix} = \left[\frac{\partial \heartsuit}{\partial \xi}, \ \frac{\partial \heartsuit}{\partial \eta} \right] \left[\widetilde{J} \right] \ . \tag{6.35}$$

The derivatives are arranged in row matrices because these objects are *gradients* of the $\heartsuit$ function [compare with (5.17)]. The matrix

$$[\tilde{J}] = \begin{bmatrix} \dfrac{\partial \xi}{\partial x} & \dfrac{\partial \xi}{\partial y} \\[2mm] \dfrac{\partial \eta}{\partial x} & \dfrac{\partial \eta}{\partial y} \end{bmatrix} , \qquad (6.36)$$

is the **Jacobian matrix** of the mapping $\xi = \xi(x,y), \eta = \eta(x,y)$, which is the inverse of the map $x = x(\xi,\eta), y = y(\xi,\eta)$ of Eq. (6.17). The question is how to evaluate the partial derivatives of the type $\partial \xi / \partial x$, since the inverse of the map (6.17) is not known (at least not in general). If we start the chain rule from the other end (switching the role of the variables), we obtain

$$\begin{bmatrix} \dfrac{\partial \heartsuit}{\partial \xi}, & \dfrac{\partial \heartsuit}{\partial \eta} \end{bmatrix} = \begin{bmatrix} \dfrac{\partial \heartsuit}{\partial x}, & \dfrac{\partial \heartsuit}{\partial y} \end{bmatrix} \begin{bmatrix} \dfrac{\partial x}{\partial \xi} & \dfrac{\partial x}{\partial \eta} \\[2mm] \dfrac{\partial y}{\partial \xi} & \dfrac{\partial y}{\partial \eta} \end{bmatrix} , \qquad (6.37)$$

and inverting the Jacobian matrix in equation (6.35) we get

$$\begin{bmatrix} \dfrac{\partial \heartsuit}{\partial \xi}, & \dfrac{\partial \heartsuit}{\partial \eta} \end{bmatrix} = \begin{bmatrix} \dfrac{\partial \heartsuit}{\partial x}, & \dfrac{\partial \heartsuit}{\partial y} \end{bmatrix} [\tilde{J}]^{-1} . \qquad (6.38)$$

Comparing (6.37) and (6.38) yields

$$[J] = \begin{bmatrix} \dfrac{\partial x}{\partial \xi} & \dfrac{\partial x}{\partial \eta} \\[2mm] \dfrac{\partial y}{\partial \xi} & \dfrac{\partial y}{\partial \eta} \end{bmatrix} = [\tilde{J}]^{-1} , \qquad (6.39)$$

where $[J]$ is the **Jacobian matrix** of the map (6.17). The elements of $[J]$ are directly available from the matrix in (6.17). However, even more useful is to start from (6.16), and by definition the Jacobian matrix is then

$$[J] = \begin{bmatrix} \displaystyle\sum_{i=1}^{3} \dfrac{\partial N_i}{\partial \xi} x_i , & \displaystyle\sum_{i=1}^{3} \dfrac{\partial N_i}{\partial \eta} x_i \\[4mm] \displaystyle\sum_{i=1}^{3} \dfrac{\partial N_i}{\partial \xi} y_i , & \displaystyle\sum_{i=1}^{3} \dfrac{\partial N_i}{\partial \eta} y_i \end{bmatrix} . \qquad (6.40)$$

Note that the Jacobian matrix may be expressed as the product of two matrices:

$$[J] = [\mathbf{x}]^T [\text{Nder}] , \qquad (6.41)$$

where [x] collects the coordinates of the nodes (three nodes, for the triangle)

$$[\mathbf{x}] = \begin{bmatrix} x_1 \ , \ y_1 \\ x_2 \ , \ y_2 \\ x_3 \ , \ y_3 \end{bmatrix} \ , \tag{6.42}$$

and [Nder] collects in each row the gradient of the basis function with respect to the parametric coordinates

$$[\mathtt{Nder}] = \begin{bmatrix} \dfrac{\partial N_1}{\partial \xi} \ , \ \dfrac{\partial N_1}{\partial \eta} \\[2mm] \dfrac{\partial N_2}{\partial \xi} \ , \ \dfrac{\partial N_2}{\partial \eta} \\[2mm] \dfrac{\partial N_3}{\partial \xi} \ , \ \dfrac{\partial N_3}{\partial \eta} \end{bmatrix} \ . \tag{6.43}$$

The calculation of the spatial derivatives by an isoparametric geometric cell (recall that the finite elements in SOFEA encapsulate the calculation of basis functions and their derivatives in the gcell class) is a straightforward rewrite of the above formulas. The method bfundsp takes three arguments: a descendent of the class gcell (the objects on which a method is being invoked are by convention called self in SOFEA), and the two arrays (6.43) and (6.42). The dimensions of the two arrays are (line 0013): nbfuns= number of basis functions (= 3 for the triangle), and dim= number of space dimensions (= 2 for the triangle).

```
0013 function Ndersp = bfundsp¹ (self, Nder, x)
0014      [nbfuns,dim] = size(Nder);
0015      if (size(Nder) ~= size(x))
0016          error('Wrong dimensions of arguments!');
0017      end
```

The Matlab code on line 0018 is literally the formula (6.41).

```
0018      J = x' * Nder;% Compute the Jacobian matrix
0019      detJ = det(J);% Compute the Jacobian
```

The Jacobian (determinant of the Jacobian matrix) should be positive. An error is reported when the Jacobian is non-positive.

```
0020      if (detJ <= 0) % trouble
0021          error('Non-positive Jacobian');
```

---

[1] Folder: SOFEA/classes/gcell/@gcell

```
0022      end
```

The generic case is treated in line 0023, which transcribes equation (6.35) for each basis function at the same time by working with matrices: each row is the gradient of one basis function. Therefore, the result for the triangle is

$$[\text{Ndersp}] = \begin{bmatrix} \dfrac{\partial N_1}{\partial x} , & \dfrac{\partial N_1}{\partial y} \\[2ex] \dfrac{\partial N_2}{\partial x} , & \dfrac{\partial N_2}{\partial y} \\[2ex] \dfrac{\partial N_3}{\partial x} , & \dfrac{\partial N_3}{\partial y} \end{bmatrix} . \tag{6.44}$$

```
0023      Ndersp = Nder * inv(J);
0024 end
```

To round off the discussion in this section, we need to present the code that evaluates the basis functions (6.12–6.14) and the derivatives of the basis functions with respect to the parametric coordinates ξ, η. For the linear triangle T3 (class gcell_T3) the two methods are delightfully simple: the method **bfun** computes a column array of the basis function values, N_j in row j, given the parametric coordinates $\xi \leftarrow$ param_coords(1), $\eta \leftarrow$ param_coords(2).

```
0008 function val = bfun²(self,param_coords)
0009      val = [(1 - param_coords(1) - param_coords(2));...
0010          param_coords(1); ...
0011          param_coords(2)];
0012      return;
0013 end
```

The method **bfundpar** returns the array (6.43) with three rows (one for each basis function), with the gradient of the basis function j with respect to ξ, η in row j.

```
0010 function val = bfundpar³(self, param_coords)
0011      val = [-1 -1; ...
0012          +1  0; ...
0013           0 +1];
```

---

[2]Folder: SOFEA/classes/gcell/@gcell_T3
[3]Folder: SOFEA/classes/gcell/@gcell_T3

```
0014      return;
0015 end
```

6.7   Numerical integration

Treading on the stepping stones of the discussion in Section 2.9, we formulate the numerical integration procedure for the linear triangle. We begin by highlighting the role of the Jacobian matrix.

Consider a map from ξ, η to x, y: a slight generalization of (6.16) in that the map is not necessarily linear (see Fig. 6.9)

$$\begin{bmatrix} x \\ y \end{bmatrix} = \begin{bmatrix} x(\xi, \eta) \\ y(\xi, \eta) \end{bmatrix} . \tag{6.45}$$

The parallelogram (rectangle) generated by the vectors $[d\xi, 0]^T$ and $[0, d\eta]^T$ (given in components in the Cartesian coordinate system ξ, η), has the area of ($\times$ is the cross product symbol)

$$\begin{bmatrix} d\xi \\ 0 \end{bmatrix} \times \begin{bmatrix} 0 \\ d\eta \end{bmatrix} = d\xi d\eta .$$

Remember, we are talking two dimensions: the cross product is a scalar.

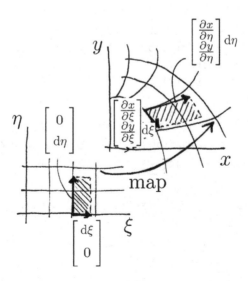

Fig. 6.9   Mapping of areas for a general map between coordinates.

The two vectors $[\mathrm{d}\xi, 0]^T$ and $[0, \mathrm{d}\eta]^T$ are mapped by the map (6.45) to vectors

$$
\begin{bmatrix} \mathrm{d}\xi \\ 0 \end{bmatrix} = \mathrm{d}\xi \begin{bmatrix} 1 \\ 0 \end{bmatrix} \longrightarrow \mathrm{d}\xi \begin{bmatrix} \dfrac{\partial x}{\partial \xi} \\[2mm] \dfrac{\partial y}{\partial \xi} \end{bmatrix} , \quad \begin{bmatrix} 0 \\ \mathrm{d}\eta \end{bmatrix} = \mathrm{d}\eta \begin{bmatrix} 0 \\ 1 \end{bmatrix} \longrightarrow \mathrm{d}\eta \begin{bmatrix} \dfrac{\partial x}{\partial \eta} \\[2mm] \dfrac{\partial y}{\partial \eta} \end{bmatrix} ,
$$

$$(6.46)$$

where the square brackets hold components in the standard Cartesian basis. Note that these vectors are **tangent** to the coordinate curves, which consist of the points in the physical space x, y that are maps of the curves $\xi = \mathrm{const}$ and $\eta = \mathrm{const}$. The area of the hatched parallelogram in Fig. 6.9 is

$$
\mathrm{d}\xi \begin{bmatrix} \dfrac{\partial x}{\partial \xi} \\[2mm] \dfrac{\partial y}{\partial \xi} \end{bmatrix} \times \mathrm{d}\eta \begin{bmatrix} \dfrac{\partial x}{\partial \eta} \\[2mm] \dfrac{\partial y}{\partial \eta} \end{bmatrix} = \mathrm{d}\xi \mathrm{d}\eta \begin{bmatrix} \dfrac{\partial x}{\partial \xi} \\[2mm] \dfrac{\partial y}{\partial \xi} \end{bmatrix} \times \begin{bmatrix} \dfrac{\partial x}{\partial \eta} \\[2mm] \dfrac{\partial y}{\partial \eta} \end{bmatrix} . \qquad (6.47)
$$

Compare this equation with (6.39): the two vectors in the cross product are the columns of the Jacobian matrix from (6.39). In fact, the cross product of the columns is the determinant of the Jacobian matrix (or, as the determinant is known, the Jacobian). Therefore, the map (6.45) maps areas as

$$
\mathrm{d}\xi \mathrm{d}\eta \longrightarrow \mathrm{d}\xi \mathrm{d}\eta \det [J] . \qquad (6.48)
$$

As a consequence of (6.48), we have the following change of coordinates in integrals:

$$
\int_{S_{[x,y]}} f(x,y)\mathrm{d}x\mathrm{d}y = \int_{S_{[\xi,\eta]}} f(\xi,\eta) \det [J(\xi,\eta)] \,\mathrm{d}\xi\mathrm{d}\eta . \qquad (6.49)
$$

Numerical quadrature rules take advantage of the relative ease with which these rules may be formulated on standard shapes, triangles, squares, cubes, etc. Thus, the integral on the left of (6.49) will be approximated as

$$
\int_{S_{[x,y]}} f(x,y)\mathrm{d}x\mathrm{d}y \approx \sum_{k=1}^{M} f(\xi_k, \eta_k) \det [J(\xi_k, \eta_k)] \, W_k . \qquad (6.50)
$$

In the `SOFEA` toolbox, the surface Jacobian $\det [J(\xi,\eta)]$ is computed for two-dimensional manifold geometric cells by the method `Jacobian_surface`; it is discussed in Section 9.4.

Table 6.1 Integration rules on the standard triangle; $a = 0.816847572980459$, $b = 0.091576213509771$, $c = 0.108103018168070$, $d = 0.445948490915965$

Rule	Coordinates ξ_j, η_j	Weights W_j	Integrates exactly
1-point	1/3, 1/3	1/2	linear polynomial
3-point	2/3, 1/6	1/6	quadratic polyn.
	1/6, 2/3	1/6	
	1/6, 1/6	1/6	
6-point	a, b	0.109951743655322	quartic polyn.
	b, a	0.109951743655322	
	b, b	0.109951743655322	
	c, d	0.223381589678011	
	d, c	0.223381589678011	
	c, c	0.223381589678011	

We will introduce three integration rules for the standard triangle, one-point, three-point, and six-point quadrature, but many other rules are available: a number of authors have compiled tables, see for instance Hughes' book [Hughes (2000)]. The 1-point rule will be able to integrate linear polynomials in ξ, η exactly, and the 3-point does the job for up to quadratic polynomials in ξ, η. The six-point rule is good for fourth order polynomials, which may seem an overkill for applications with linear triangles, but its worth will be appreciated later. Table 6.1 gives the coordinates of the integration points, and their weights.

6.8   Conductivity matrix

As discussed already in Chapter 3, all the problem dependent code is concentrated in a descendent of the `feblock` class. In particular, the two-dimensional heat diffusion model of this chapter is implemented in the `feblock_diffusion` finite element block class.

The conductivity (6.24) and other matrices are computed by evaluating the contributions from each element separately, storing these contributions element-by-element in a cell array, and then finally assembling all the element contributions into the overall system matrix. Therefore, the conductivity matrix would be computed element-by-element as

$$K_{ji} = \sum_e \int_{\triangleright_e} (\text{grad}N_j) \, \boldsymbol{\kappa}(\text{grad}N_i)^T \, \Delta z \, \mathrm{d}S \ . \tag{6.51}$$

The procedure is illustrated in Fig. 6.10: Only the gradients of the basis functions A, B, C are nonzero over the triangle e. The conductivity matrix *of this element* is calculated as if there were only three nodes in the whole mesh, all with free degrees of freedom (temperatures). It is therefore a 3×3 matrix K_e. However, only the unknown degrees of freedom are given equation numbers on the global level. In this example, we assume that node A carries the unknown number 13, node C carries the unknown 61, and node B is associated with prescribed temperature, and correspondingly a zero (0) indicates that there is no equation for node B. The K_e matrix should be added to the global matrix as indicated in the sum (6.51), and the arrows pointing towards the appropriate elements $((13, 13), (13, 61), (61, 13), (61, 61))$ of the global conductivity matrix K indicate the so-called *assembly* of the element matrix. This assembly process is executed in the assemble[4] method of the classes dense_sysmat and sparse_sysmat.

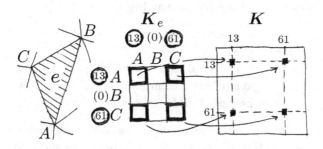

Fig. 6.10   Assembly of the element conductivity matrix.

The method conductivity returns an array (ems) of element matrix objects (class elemat), each of which represents the conductivity matrix of a single element. The method begins by retrieving some information from the parent class, such as gcells (cell array of the geometric cells), integration_rule, and the material mat.

```
0009 function ems = conductivity⁵(self, geom, theta)
0010     gcells = get(self.feblock,'gcells');
0011     ngcells = length(gcells);
0012     nfens = get(gcells(1),'nfens');
0013     dim = get(geom,'dim');
0014     % Pre-allocate the element matrices
0015     ems(1:ngcells) = deal(elemat);
```

---

[4]Folder: SOFEA/classes/sysmat/@dense_sysmat
[5]Folder: SOFEA/classes/feblock/@feblock_diffusion

```
0016    % Integration rule
0017    integration_rule = get(self.feblock, 'integration_rule');
0018    pc = get(integration_rule, 'param_coords');
0019    w  = get(integration_rule, 'weights');
0020    npts_per_gcell = get(integration_rule, 'npts');
0021    % Material
0022    mat = get(self.feblock, 'mater');
0023    kappa = get(get(mat,'property'),'conductivity');
```

The loop over all the geometric cells starts with the retrieval of the connectivity (i.e. the numbers of the nodes which are connected together by the cell), and of the array of the node coordinates, x (compare with (6.42)). Then, the element conductivity matrix Ke is initialized to zero, and the loop over all the quadrature points may begin.

```
0024    % Now loop over all gcells in the block
0025    for i=1:ngcells
0026        conn = get(gcells(i), 'conn'); % connectivity
0027        x = gather(geom,conn,'values','noreshape');%coord
0028        Ke = zeros(nfens); % element matrix
```

The loop over the integration points begins with the calculation of the basis functions and of the gradients of the basis functions with respect to the parametric coordinates; see array (6.43), followed by the computation of the spatial derivatives of the basis functions, Nspder, and the Jacobian, detJ. Note that the method used, Jacobian_volume, computes a *volume* Jacobian: even though the method works with representation of temperatures as functions of two variables, x, y, the problem that is being solved is still the heat conduction through a three-dimensional solid (there is no such thing as a two-dimensional body for an engineer; only mathematicians seem to know where to find them).

```
0029        % Loop over all integration points
0030        for j=1:npts_per_gcell
0031            N = bfun (gcells(i), pc(j,:));
0032            Nder = bfundpar (gcells(i), pc(j,:));
0033            Nspder = bfundsp(gcells(i), Nder, x);
0034            detJ = Jacobian_volume(gcells(i),pc(j,:),x);
```

Often for orthotropic materials the axes of orthotropy vary from point-to-point. In that case it makes sense to describe the material properties

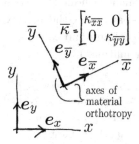

Fig. 6.11   Directions of material orthotropy.

in local Cartesian coordinates, and then allow the finite element block to define a transformation matrix between the local coordinate directions and the global Cartesian basis: refer to Fig. 6.11. The attribute of the material property object is thus the material conductivity in the *local basis*, $e_{\overline{x}}$, $e_{\overline{y}}$

$$[\overline{\kappa}] = \begin{bmatrix} \kappa_{\overline{x}} & 0 \\ 0 & \kappa_{\overline{y}} \end{bmatrix} , \tag{6.52}$$

which is rotated into the global Cartesian basis using the transformation matrix (rotation matrix)

$$[R_m] = \big[\, [e_{\overline{x}}]\ [e_{\overline{y}}]\,\big] . \tag{6.53}$$

The columns of $[R_m]$ are the components of the basis vectors $e_{\overline{x}}$, $e_{\overline{y}}$ in the global Cartesian coordinates. The material conductivity matrix in the global basis is then expressed using the ordinary transformation rule

$$[\kappa] = [R_m][\overline{\kappa}][R_m]^T . \tag{6.54}$$

The finite element block computes the local material directions using either a user-supplied orientation matrix, or the default identity matrix.

```
0035            Rm = material_directions(self,
                       gcells(i),pc(j,:),x);
```

Now we exercise the integration rule. Note that the element conductivity matrix is computed in one shot, since the gradients of the basis functions are arranged as rows of the Nspder matrix.

```
0036            Ke=Ke+Nspder*Rm*kappa*Rm'*Nspder'*detJ*w(j);
```

Finally, the computed element conductivity matrix `Ke` is stored in the `ems(i)` object of class `elemat`, both the matrix itself and the equation numbers that go with each column and each row (compare with Fig. 6.10: 13, 0, 61; zero means "no equation").

```
0037          end
0038          ems(i)=set(ems(i), 'mat', Ke);
0039          ems(i)=set(ems(i), 'eqnums',
                        gather(temp,conn,'eqnums'));
0040      end
0041      return;
0042 end
```

Since the topic of the Jacobians has been brought up, we point out how the volume Jacobian is computed for the triangle element T3. We ignore everything that does not pertain to the present case, and we focus on line 0021: the volume Jacobian is computed as the product of the surface Jacobian (determinant of (6.39)) and the "other dimension" (thickness).

```
0015 function detJ = Jacobian_volume⁶(self, pc, x)
0016  ...
0021          detJ=Jacobian_surface(self,pc,x)*
                        other_dimension(self,pc,x);
0022      end
0023 end
```

6.9   Surface heat transfer matrix and load

In the preceding section we have been dealing with volume integrals, evaluated over the area S_c which together with the thickness Δz gave the volume. The surface heat transfer matrix (6.29) and the surface heat transfer load (6.28) (and also the prescribed heat flux load (6.27)) require integration over the bounding surface, evaluated over the curves, $C_{c,3}$ (or $C_{c,2}$), which together with the thickness gives a surface area. As the volume integrals are evaluated over the area of the triangles in the mesh, the surface integrals will be computed over the edges of these triangles.

Evaluating the basis functions (6.12–6.14) along the edges of the standard triangle, we may observe that the basis function associated with the

---

[6]Folder: `SOFEA/classes/gcell/@gcell_2_manifold`

opposite vertex is identically zero, and the other two at the nodes at the end-points of the edge vary linearly along the edge. In fact, completely in agreement with the basis functions (2.22) defined on the line element L2. Therefore, integrating an expression along the edge of the triangle T3 that connects nodes i, j yields exactly the same result as integrating along the line element L2 that connects nodes i, j. However, the two approaches are rather different in terms of implementation: if, for the purpose of numerical integration, we use the element L2, the design of the numerical integration code will be reusable: the same piece of code may be used to integrate quantities along a curve which is tiled with finite element edges with linear variation of basis functions – the triangle T3, the quadrilateral Q4, the hexahedron H8, the tetrahedron T4. (These elements will be discussed shortly.)

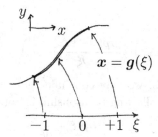

Fig. 6.12   Mapping of the standard interval to a Cartesian space.

Let us continue with the discussion of the curve integrals. The goal is to evaluate

$$\int_C f(\boldsymbol{p}) \, \mathrm{d}C \,, \tag{6.55}$$

where we will assume that the curve C may be "embedded" in a three-dimensional, two-dimensional, or one-dimensional Euclidean space (i.e. it may be a spatial curve, plane curve, or just an interval on the real line). Correspondingly, the point $\boldsymbol{p}$ on the curve C will have an appropriate number of components, three, two, or one.

To perform the integral, the elementary length $\mathrm{d}C$ is needed. The point $\boldsymbol{p}$ on the curve will be assumed to be the result of the mapping of the standard interval $-1 \le \xi \le +1$ (compare with the 1-D map (2.25), and refer to Fig. 6.12, where the map is two-dimensional)

$$p = g(\xi) \, . \tag{6.56}$$

For two closely spaced points on the curve, $p(\xi)$ and $p(\xi + \Delta\xi)$, where $\Delta\xi$ is the distance between the two points in the standard interval, the second point may be obtained from the first using the first two terms of the Taylor series as

$$p(\xi + \Delta\xi) = p(\xi) + \frac{\partial p(\xi + \varepsilon\Delta\xi)}{\partial \xi}\Delta\xi \, , \qquad 0 \le \varepsilon \le 1. \tag{6.57}$$

The two points may be connected with a vector approximately tracking the curve (see Fig. 6.13),

$$p(\xi + \Delta\xi) - p(\xi) = \frac{\partial p(\xi + \varepsilon\Delta\xi)}{\partial \xi}\Delta\xi \, ,$$

whose length (squared) is

$$(\Delta C)^2 = \left(\frac{\partial p(\xi + \varepsilon\Delta\xi)}{\partial \xi}\Delta\xi\right) \cdot \left(\frac{\partial p(\xi + \varepsilon\Delta\xi)}{\partial \xi}\Delta\xi\right) = \left\|\frac{\partial p(\xi + \varepsilon\Delta\xi)}{\partial \xi}\right\|^2 (\Delta\xi)^2 \, .$$

Skipping over the details, we may conclude that for infinitesimally short intervals, $\Delta\xi \to d\xi$, the following relationship is obtained

$$dC = \left\|\frac{\partial p(\xi)}{\partial \xi}\right\| d\xi \, , \tag{6.58}$$

where $\dfrac{\partial p(\xi)}{\partial \xi}$ is the vector ***tangent*** to the curve at ξ, and $\left\|\dfrac{\partial p(\xi)}{\partial \xi}\right\|$ is the Jacobian to be used in the change-of-variables device for the curve integral. The above developments should be compared with Eq. (2.26) and the discussion preceding it: indeed, the current result is only a slight generalization.

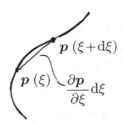

Fig. 6.13   Length of a curve.

Let us now specialize these developments to the L2 element: the map (6.56) reads

$$p = g(\xi) = \sum_{i=1}^{2} N_i(\xi) x_i \ .$$ (6.59)

where the N_i's are given by (2.27). Therefore, the tangent vector (see (6.58)) reads

$$\frac{\partial p(\xi)}{\partial \xi} = \frac{x_2 - x_1}{2} \ ,$$

and the Jacobian is $h/2$, where $h = ||x_2 - x_1||$ is the length of the element.

Of course, for general elements with n nodes, the implementation computes the tangent as

$$\frac{\partial p(\xi)}{\partial \xi} = \texttt{x'*Nder} \ ,$$ (6.60)

using the following two matrices,

$$[\texttt{x}] = \begin{bmatrix} x_1 \ , \ y_1 \\ x_2 \ , \ y_2 \\ ..., \ ... \\ x_n \ , \ y_n \end{bmatrix} \ ,$$ (6.61)

where the number of columns is equal to the number of spatial dimensions, 1, 2 (which is assumed in (6.61)), or 3, and [Nder] collects in each row the gradient of the basis function with respect to the parametric coordinate

$$[\texttt{Nder}] = \begin{bmatrix} \dfrac{\partial N_1}{\partial \xi} \\[2mm] \dfrac{\partial N_2}{\partial \xi} \\[2mm] ... \\[2mm] \dfrac{\partial N_n}{\partial \xi} \end{bmatrix} \ ,$$ (6.62)

These matrices should be compared with those defined for the triangle T3, Eqs. (6.42) and (6.43). The only difference is the number of space dimensions, the number of basis functions, and the number of parametric dimensions; all of these are taken into account by the Matlab code automatically.

For the one-dimensional manifold geometric cell, this computation is implemented in the method Jacobian_curve of the class gcell_1_manifold. Compare formula (6.60) with line 0015.

```
0013 function detJ = Jacobian_curve⁷(self, pc, x)
0014     Nder = bfundpar (self, pc);
0015     tangent =x'*Nder;
0016     [sdim, ntan] = size(tangent);
0017     if    ntan==1 % 1-D gcell
0018         detJ = norm(tangent);
0019     else
0020         error('Got an incorrect size of tangent');
0021     end
0022 end
```

Now that we have worked out how to integrate along a curve, we may come back to our original goal. The method surface_transfer computes the heat transfer matrix (6.29) by integrating over the appropriate part of the surface. Note that the surface needs to be discretized by compatible geometric cells: when the domain (volume) is covered by three-node triangles, the boundary (surface) is covered by two-node line segments. Most of the preparation steps are straightforward, and are therefore omitted. The loop over the geometric cells computes the heat transfer matrix for each element. The Jacobian is computed in a way that is appropriate for the geometric cells (yet another example of dynamic method dispatch): the method Jacobian_surface.

```
0009 function ems = surface_transfer⁸(self, geom, temp)
...
0021     h = self.surface_transfer;
0022     % Now loop over all gcells in the block
0023     for i=1:ngcells
0024         conn = get(gcells(i), 'conn'); % connectivity
0025         x = gather(geom, conn, 'values', 'noreshape');
0026         He = zeros(nfens); % element matrix
0027         for j=1:npts_per_gcell % Loop over integr. points
0028             N = bfun(gcells(i), pc(j,:));
0029             detJ = Jacobian_surface(gcells(i),pc(j,:),x);
```

---

[7] Folder: SOFEA/classes/gcell/@gcell_1_manifold
[8] Folder: SOFEA/classes/feblock/@feblock_diffusion

```
0030                He = He + h*N*N' * detJ * w(j);
0031      end
0032      ems(i)=set(ems(i),'mat', He);
0033      ems(i)=set(ems(i),'eqnums',
                          gather(temp,conn,'eqnums'));
0034    end
0035    return;
0036 end
```

The surface Jacobian is computed for the L2 (line) element following the same principle as before for the volume Jacobian for the element T3. Focusing just on this case, we see that the surface Jacobian is computed as the product of the curve Jacobian and the "other dimension" (thickness).

```
0015 function detJ = Jacobian_surface⁹(self, pc, x)
0016 ...
0021      detJ=Jacobian_curve(self,pc,x)*
                          other_dimension(self,pc,x);
0022    end
0023 end
```

Computation of the surface transfer loads could duplicate the work done in the previous method (the same kind of integral), but in order to avoid this duplication, we use the following trick: compute an array of the element heat surface transfer matrices, and multiply by the vector of ambient temperatures (whenever they are nonzero); that gives us the vector of loads. The method `surface_transfer_loads` is therefore quite straightforward: compute the element surface heat transfer matrices (line 0010), then loop over the geometric cells and retrieve the ambient temperatures from the field `amb`. Provided this vector is nonzero, the product `He*pT` is stored in the element vector object.

```
0009 function evs=surface_transfer_loads¹⁰(self,geom,temp,amb)
0010    ems = surface_
                    transfer(self,geom,temp);
0012    gcells = get(self,feblock, matrices
0013    evs(1:length(gcells)) = deal(elevec);
0014    % Now loop over all gcells in the block
0015    for i=1:length(gcells)
```

---

⁹Folder: SOFEA/classes/gcell/@gcell_1_manifold
¹⁰Folder: SOFEA/classes/feblock/@feblock_diffusion

```
0016              conn = get(gcells(i), 'conn'); % connectivity
0017              pT = gather(amb, conn, 'prescribed_values');
0018              if norm (pT) ~= 0
0019                  He =  get(ems(i),'mat'); % element matrix
0020                  evs(i) = set(evs(i), 'vec', He*pT);
0021                  evs(i) = set(evs(i), 'eqnums', ...
0022                      gather(temp, conn, 'eqnums'));
0023              end
0024          end
0025      return;
0026 end
```

Exercises

(1) For the mesh shown in Figure 6.14

  (a) calculate N_7 at the following points (given by $[x, y]$): $[0, 0]$, $[4, 3]$, $[2, 1.5]$, $[5, 6]$, $[4/3, 3]$.

  (b) Evaluate the basis function N_4 at the midpoint of all the edges connected to node 7.

  (c) Evaluate the basis function N_2 at the barycenter of all the triangles in the mesh.

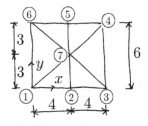

Fig. 6.14   Mesh.

(2) Same data as in (1). Paper and pencil exercise.

  (a) Calculate the degrees of freedom T_i at all the nodes so that the function $T(x, y) = \sum_{i=1}^{7} N_i(x, y)T_i$ interpolates the function $f(x, y) = 2x + 3 - 5y$ at the nodes.

  (b) Plot the error $|f(x, y) - T(x, y)|$ on the mesh.

(3) Pencil and paper exercise.

    (a) For the mesh shown in Fig. 6.14 calculate the derivatives of the basis function N_7 with respect to x at the following points (given by $[x, y]$): $[0, 0]$, $[4, 3]$, $[2, 1.5]$, $[5, 6]$, $[4/3, 3]$. Warning: the derivatives may be multi-valued (different value in each triangle that shares the point), in which case indicate the value and to which triangle it belongs).

    (b) For the mesh shown in Fig. 6.14 calculate the derivatives of the basis function N_7 with respect to y at the following points (given by $[x, y]$): $[0, 0]$, $[4, 3]$, $[2, 1.5]$, $[5, 6]$, $[4/3, 3]$. Warning: the derivatives may be multi-valued (different value in each triangle that shares the point), in which case indicate the value and to which triangle it belongs).

    (c) For the mesh shown in Fig. 6.14 calculate the derivatives of the basis function N_2 with respect to x within all the triangles of the mesh. Produce a drawing to illustrate the derivative on the mesh as a surface $z = s(x, y)$.

    (d) For the mesh shown in Figure 6.14 calculate the derivatives of the basis function N_4 with respect to y within all the triangles of the mesh. Produce a drawing to illustrate the derivative on the mesh as a surface $z = s(x, y)$.

(4) Solve a steady-state heat conduction problem in the domain of Fig. 6.15. This is a pencil and paper exercise. The thermal conductivity coefficient of the material is $\kappa = 0.2$ W/m/K°. Temperature is prescribed on the edges A, B as $\overline{T} = (300 + xy)$ K° (x and y are measured from the center of the rectangle). Surface heat transfer is prescribed through edge D: heat surface transfer coefficient $h = 5$ W/m²/K°, ambient temperature $T_a = 290$K°. The thickness is $\Delta z = 1$.

    (a) Use the two two-element meshes shown in the figure on the right. Compute the temperature at the upper right-hand corner of the domain and plot the temperature as a surface.

(5) Repeat the procedure of Section 6.2 of the textbook to derive a one-dimensional analog of Eq. (6.11) to correspond to heat flow through the thickness of a plate (or a wall) (see Fig. 6.16): only the x-component of the heat flux will be non-zero, and the trial and test functions will be functions of x (and in the case of the trial function also of time). This is a pencil and paper exercise.

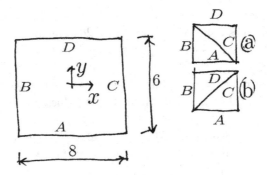

Fig. 6.15   Definition of the domain and the meshes.

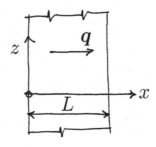

Fig. 6.16   One-dimensional heat conduction problem.

Chapter 7

Steady-state Heat Conduction Solutions

The ordinary differential equations that result from the discretization in space, Eqs. (6.34), lead to steady-state solutions when $\partial T_i(t)/\partial t = 0$, and $\partial \overline{T}_i(t)/\partial t = 0$. The latter condition is necessary, while the former follows when all the transients in the solution decay (in infinite time, in general).

7.1 Steady-state heat conduction equation

Substituting the vanishing temperature rates into (6.34), we obtain

$$\sum_{\text{free } i} K_{ji}T_i + \sum_{\text{free } i} H_{ji}T_i = L_{\overline{K},j} + L_{\overline{H},j} + L_{Q,j} + L_{q2,j} + L_{q3,j} \quad \forall \text{ free } j,$$

(7.1)

which is a system of linear equations for the unknown nodal temperatures. The nodal temperatures are now just numbers, not functions. Let us look at a few examples of steady-state heat conduction.

7.2 Thick-walled tube

The first example is a thick-walled rectangular tube, with the outside temperature being prescribed as zero, and the interior surface (perfectly) insulated. The material is isotropic. As shown in Fig. 7.1, the planes of symmetry may be used to reduce the size of the problem. Therefore, only one quarter is discretized, and perfect insulation is applied at the symmetry planes (no heat flows through the symmetry planes). There is a distributed heat source in the material (for instance, such as heat released by curing cement paste).

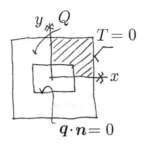

Fig. 7.1   Heat diffusion in a thick-walled rectangular tube.

The Matlab script is `lshape1`[1]. The first few lines define some ancillary variables, and then the two-dimensional mesh generator of triangle meshes is invoked. The generator is thoroughly described in the user's guide `targe2_users_guide.pdf`[2], but we will say a few words about its Matlab interface. The first argument is a cell array, each element a string (character array), with one command for the mesh generator. Thus, the first six strings define the curves that bound the domain (straight-line segments), the line 0011 defines a subregion (chunk of area to be covered with triangles), and the last line defines the mesh size. The second argument to `targe2_mesher` is the thickness of the slab (default value of 1.0).

```
0001 kappa=[0.2 0; 0 0.2]; % conductivity matrix
0002 Q=0.01100; % uniform heat source
0003 num_integ_pts=1; % 1-point quadrature
0004 [fens,gcells] = targe2_mesher({...
0005         ['curve 1 line 20 0 48 0'],...
0006         ['curve 2 line 48 0 48 48'],...
0007         ['curve 3 line 48 48 0 48'],...
0008         ['curve 4 line 0 48 0 13'],...
0009         ['curve 5 line 0 13 20 13'],...
0010         ['curve 6 line 20 13 20 0'],...
0011         'subregion 1 property 1 boundary 1 2 3 4 5 6',...
0012         ['m-ctl-point constant 3.5']
0013         }, 1.0);
```

Next, the property object appropriate for the heat diffusion model is created, `property_diffusion`, and supplied the material conductivity matrix

---

[1]Folder: `SOFEA/examples/heat_diffusion`
[2]Folder: `SOFEA/meshing/targe2`

κ, and the heat source Q. The material object acts as a mediator between the property object and the finite element block. The finite element block of class `feblock_diffusion` is created, with attributes: the material, the array of geometric cells, and an integration rule (class `tri_rule` is used for triangles).

```
0014 prop=property_diffusion(
        struct('conductivity',kappa,'source',Q));
0015 mater=mater_diffusion (struct('property',prop));
0016 feb = feblock_diffusion (struct ('mater',mater,...
0017     'gcells',gcells,...
0018     'integration_rule',tri_rule(num_integ_pts)));
```

Two fields are created: `geom` represents the geometry (i.e. the locations of the nodes), and it is therefore initialized from the finite element node array, `fens`; and `theta` represents the temperatures at the nodes, and it is initially undefined, except for the number of nodes `nfens`.

```
0019 geom =field(struct('name',['geom'],'dim',2,'fens',fens));
0020 theta=field(struct('name',['theta'],'dim',1,'nfens',...
0021     get(geom,'nfens')));
```

The essential boundary conditions are next applied to the temperature field. The utility function `fenode_select` is used to select nodes from the `fens` array based on their location: nodes which fall into given bounding boxes are selected ($[x_{lo} \; x_{hi} \; y_{lo} \; y_{hi}] = [48 \; 48 \; 0 \; 48]$ and so on for the other box); to avoid problems with testing whether of point is inside or outside a box of zero "thickness", the boxes are for the purpose of the "in"-test inflated by 0.01. The array `prescribed` is filled with ones to indicate that all degrees of freedom are to be prescribed, the components to be prescribed are passed as an empty array (line 0026), which simply means all components are affected. The values to which the temperatures are being prescribed are all zeros. The data defining the essential boundary conditions are set in the field (`set_ebc`), and then applied (method `apply_ebc` on line 0029). The free node parameters are then assigned global equation numbers with the method `numbereqns`.

```
0022 fenids=[fenode_select(fens,struct('box',[48 48 0 48],...
0023     'inflate', 0.01)),...
0024     fenode_select(fens,struct('box',[0 48 48 48],...
0025     'inflate', 0.01))];
```

```
0026 prescribed=ones(length(fenids),1);
0027 comp=[];
0028 val=zeros(length(fenids),1);
0029 theta = set_ebc(theta, fenids, prescribed, comp, val);
0030 theta = apply_ebc (theta);
0031 theta = numbereqns (theta);
```

The conductivity matrix is sparse (the linear system to be solved is going to be moderately large, and the efficiency afforded by a sparse matrix is not to be sneezed at), and it is assembled from element conductivity matrices in line 0032. The heat load vector is assembled from element load vectors, and the solution of the linear system of equations is scattered into the theta field.

```
0032 K = start (sparse_sysmat, get(theta, 'neqns'));
0033 K = assemble (K, conductivity(feb, geom, theta));
0034 F = start (sysvec, get(theta, 'neqns'));
0035 F = assemble (F, source_loads(feb, geom, theta));
0036 theta = scatter_sysvec(theta, get(K,'mat')\get(F,'vec'));
```

The last fragment of code takes care of the graphic presentation of the results. The field colorfield holds one color (a triple of floating-point numbers) per node, and those colors are obtained from the temperature field by mapping node temperatures to colors (line 0042) using the map_data method of the data_colormap class. The geometric cells of individual finite elements are plotted twice. Once as a raised colored surface (line 0047), and the second time as a wireframe in the x, y plane (line 0049). The resulting graphic is shown in Fig. 7.2.

```
0038 gv=graphic_viewer;
0039 gv=reset (gv,[]);
0040 T=get(theta,'values');
0041 dcm=data_colormap(struct('range',[min(T),max(T)],
               'colormap',jet));
0042 colorfield=field(struct('name',['colorfield'],'data',...
0043     map_data(dcm, T)));
0044 geomT=field(struct ('name', ['geomT'], ...
0045     'data',[get(geom,'values'), get(theta,'values')]));
0046 for i=1:length (gcells)
0047     draw(gcells(i), gv, struct('x',geomT,'u',0*geomT,...
0048         'colorfield',colorfield, 'shrink',0.9));
```

```
0049     draw(gcells(i), gv, struct('x',geom, 'u',0*geom, ...
0050        'facecolor','none'));
0051 end
```

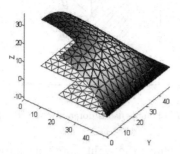

Fig. 7.2 Heat diffusion in a thick-walled rectangular tube: graphic presentation of results.

7.3 Orthotropic insert

The next example introduces nonzero essential boundary conditions, and orthotropic material properties. A square block of isotropic material is insulated on the vertical edges, and two different temperatures are applied on the horizontal edges. There is a square insert of orthotropic material within the larger square. The orientation of the material axes is indicated in Fig. 7.3. Physically the insert could be made of parallel fibers (for instance carbon), embedded in a polymer matrix. The fibers conduct heat well, while in the transverse direction the polymer matrix hampers heat conduction. The problem is solved with script **squareinsquare**[3]. The domain consists of two materials, and consequently we define two material conductivity matrices: the inner material has strongly orthotropic properties; the outer material is isotropic. The rotation matrix that defines the local material properties of the insert is set up in line 0005.

```
0001 kappainner=[2.25 0; 0 0.06]; % orthotropic conduct. matrix
0002 kappaouter=[0.25 0; 0 0.25]; % isotropic conduct. matrix
0003 alpha =-45;% local material orientation angle
0004 ca=cos(2*pi/360*alpha); sa=sin(2*pi/360*alpha);
0005 Rm = [ca, -sa;sa, ca];% local material directions
```

---

[3] Folder: SOFEA/examples/heat_diffusion

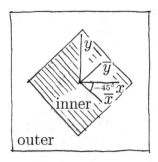

Fig. 7.3   Heat diffusion in inhomogeneous domain with orthotropic material properties.

The mesh generator defines the eight boundary segments and two subregions: note that the two subregions are assigned different numerical identifiers (1 and 2) to distinguish elements belonging to different subregions.

```
0007 [fens,gcells, groups] = targe2_mesher({...
0008          ['curve 1 line -48 -48 48 -48'],...
0009          ['curve 2 line 48 -48 48 48'],...
0010          ['curve 3 line 48 48 -48 48'],...
0011          ['curve 4 line -48 48 -48 -48'],...
0012          ['curve 5 line 0 -31 31 0'],...
0013          ['curve 6 line 31 0 0 31'],...
0014          ['curve 7 line 0 31 -31 0'],...
0015          ['curve 8 line -31 0 0 -31'],...
0016          ['subregion 1  property 1 ' ...
0017            '    boundary 1 2 3 4 -8 -7 -6 -5'],...
0018          ['subregion 2  property 2 '...
0019            '    boundary 5 6 7 8'],...
0020          ['m-ctl-point constant 4.75']
0021          }, 1.0);
```

The inner subregion consists of the geometric cells gcells(groups{2}) (groups{2} is a list of indexes of the cells that belong to the subregion 2). Note that the local material directions matrix is being supplied to the finite element block constructor (line 0027).

```
0022 propinner=property_diffusion(
          struct('conductivity',kappainner,...
0023      'source',0));
```

```
0024 materinner=mater_diffusion(struct('property',propinner));
0025 febinner = feblock_diffusion(struct ('mater',materinner,
0026     'gcells',gcells(groups{2}),...
0027     'integration_rule',tri_rule(num_integ_pts),'Rm',Rm));
0028 propouter=property_diffusion(
        struct('conductivity',kappaouter,
0029     'source',0));
0030 materouter=mater_diffusion(struct('property',propouter));
0031 febouter = feblock_diffusion(struct ('mater',materouter,...
0032     'gcells',gcells(groups{1}),...
0033     'integration_rule',tri_rule(num_integ_pts)));
```

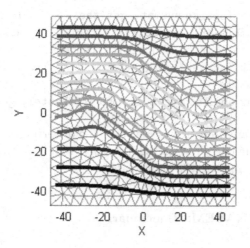

Fig. 7.4 Heat diffusion in inhomogeneous domain with orthotropic material properties: temperature distribution. Notice the distorting effect of the insert.

The boundary conditions are straightforward, but notice that the two horizontal edges are being assigned different, nonzero, temperatures.

```
0036 fenids=[fenode_select(fens,struct('box',[-48 48 -48 -48],
0037     'inflate', 0.01))];
0038 prescribed=ones(length(fenids),1);
0039 comp=[];
0040 val=zeros(length(fenids),1)+20;% cold
0041 theta = set_ebc(theta, fenids, prescribed, comp, val);
0042 fenids=[fenode_select(fens,struct('box',[-48 48 48 48],
```

```
0043    'inflate', 0.01))];
0044 prescribed=ones(length(fenids),1);
0045 comp=[];
0046 val=zeros(length(fenids),1)+57;% hot
0047 theta = set_ebc(theta, fenids, prescribed, comp, val);
0048 theta = apply_ebc (theta);
```

When assembling the conductivity matrix, the contributions from the two blocks are assembled separately. The thermal loads corresponding to nonzero essential boundary conditions (conductivity only: recall that this is steady-state) are assembled only for the outer subregion block, since there are no boundary conditions on the boundary of the inner block.

```
0050 K = start (sparse_sysmat, get(theta, 'neqns'));
0051 K = assemble (K, conductivity(febinner, geom, theta));
0052 K = assemble (K, conductivity(febouter, geom, theta));
0053 F = start (sysvec, get(theta, 'neqns'));
0054 F = assemble (F,
                nz_ebc_loads_conductivity(febouter, geom, theta));
```

The results are presented in Fig. 7.4. The distorting effect of the insert is noteworthy: in one direction the insert acts as a heat sink/source, in the perpendicular direction it is an insulator.

7.4   The T4 NAFEMS Benchmark

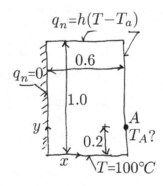

Fig. 7.5   The T4 NAFEMS benchmark geometry and boundary conditions.

This problem is one of the NAFEMS (National Agency for Finite Element Methods and Standards (UK)) benchmark tests for thermal analyses. The boundary conditions are formulated for a rectangle 0.6 meter wide by 1 meter high, with a fixed temperature of 100°C on the lower boundary, perfect insulator on the left boundary, and a heat transfer at $750\text{W/m}^2{}^\circ\text{C}$ on the other two boundaries (see Fig. 7.5). The material in the region has a thermal conductivity of $52\text{W/m}^\circ\text{C}$. The problem is to calculate the steady-state temperature distribution. A complete description of this problem is given in the paper by Cameron, Casey, and Simpson [Cameron et al. (1994)].

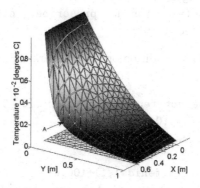

Fig. 7.6   Temperature distribution for the T4 NAFEMS benchmark.

The SOFEA solution is the Matlab script t4nafems[4]. Note well that two blocks are being created: the first for the triangular elements in the interior of the domain, and the second, edgefeb, for the edge elements (line segment elements with two nodes) along the two boundary edges of the domain with the convective boundary condition.

```
0019 edgefeb = feblock_diffusion (struct ('mater',mater,...
0020     'gcells',edge_gcells([edge_groups{[2, 3, 4]}]),...
0021     'integration_rule',gauss_rule(1,num_integ_pts),...
0022     'surface_transfer', h));
...
```

Next, we create a field to represent the prescribed ambient temperature along the boundary. The interior values are never used, only the ones on the boundary. They happen to be all equal to zero, but we will not ignore

---

[4]Folder: SOFEA/examples/heat_diffusion

them in the interest of clarity.

```
0026 amb = clone(theta, ['amb']);
0027 fenids=[
0028      fenode_select(fens,struct('box',[0.6 0.6 0 1],...
0029      'inflate', 0.01)),...
0030      fenode_select(fens,struct('box',[0 1 1 1],...
0031      'inflate', 0.01))]   ;
0032 prescribed=ones(length(fenids),1);
0033 comp=[];
0034 val=zeros(length(fenids),1)+0.0;
0035 amb = set_ebc(amb, fenids, prescribed, comp, val);
0036 amb = apply_ebc (amb);
```

The essential boundary condition on the temperature field is applied.

```
0037 fenids=[
0038      fenode_select(fens,struct('box',[0. 0.6 0 0],...
0039      'inflate', 0.01))]    ;
0040 prescribed=ones(length(fenids),1);
0041 comp=[];
0042 val=zeros(length(fenids),1)+100.0;
0043 theta = set_ebc(theta, fenids, prescribed, comp, val);
0044 theta = apply_ebc (theta);
```

The system matrix and the system load vector are assembled, including the surface heat transfer contribution (line 0047), and the surface heat transfer load (line 0051). Note that these are computed on the edge element block `edgefeb`. The contribution of the nonzero prescribed temperature is also added in.

```
0046 K = start (sparse_sysmat, get(theta, 'neqns'));
0047 K = assemble (K, conductivity(feb, geom, theta));
0048 K = assemble (K, surface_transfer(edgefeb, geom, theta));
0049 F = start (sysvec, get(theta, 'neqns'));
0050 F = assemble(F, source_loads(feb, geom, theta));
0051 F = assemble(F,
                 nz_ebc_loads_conductivity(feb, geom, theta));
0052 F = assemble(F,
                 nz_ebc_loads_surface_transfer(feb, geom, theta));
0053 F = assemble(F,
```

```
        surface_transfer_loads(edgefeb,geom,theta,amb));
```

After the solution, the temperature at the node $x = 0.6$, $y = 0.2$ (label A in Fig. 7.6), is retrieved with **gather** and printed. The calculated value of 18.2481°C agrees well with the reference solution of 18.3°C. The singularity near the corner where the two kinds of boundary conditions meet (prescribed temperature with convective surface heat transfer) is clearly visible (very large slope, that in the limit tends to infinity).

```
0054 theta = scatter_sysvec(theta, get(K,'mat')\get(F,'vec'));
0055 gather(theta,fenode_select(fens,...
0056  struct('box',[0.6 0.6 0.2 0.2],'inflate',0.01)),'values')
```

Chapter 8

Transient Heat Conduction Solutions

The ordinary differential equations (6.34) need to be numerically integrated in time as analytical solutions are not possible in general. In this chapter, we will explore a set of methods for numerical integration of the transient response of the heat conduction model.

8.1 Discretization in time for transient heat conduction

Hughes describes a finite difference method, the **generalized trapezoidal method** , including its accuracy and stability properties (Chapter 8, Reference [Hughes (2000)]). In order to unclutter the equations we will use the matrix notation, with the following definitions:

$$\widetilde{K} = [K_{ji} + H_{ji}], \quad \text{free } j, i, \tag{8.1}$$

for the effective conductivity matrix, which bundles the bulk conductivity with the surface heat transfer matrix,

$$\overline{K} = [\overline{K}_{ji} + \overline{H}_{ji}], \quad \text{free } j, \text{ prescribed } i, \tag{8.2}$$

for the rectangular conductivity matrix that relates the prescribed temperatures to the heat powers,

$$C = [C_{ji}], \quad \text{free } j, i, \tag{8.3}$$

for the capacity matrix,

$$\overline{C} = [\overline{C}_{ji}], \quad \text{free } j, \text{ prescribed } i, \tag{8.4}$$

for the rectangular capacity matrix that relates the prescribed temperature rates to the heat powers, and

$$\boldsymbol{L} = [L_{Q,j} + L_{q2,j} + L_{q3,j}], \quad \text{free } j .\tag{8.5}$$

The free temperatures and their rates are collected in column matrices

$$\boldsymbol{T} = [T_j], \quad \dot{\boldsymbol{T}} = \left[\frac{\partial T_i}{\partial t}\right], \quad \text{free } j ,\tag{8.6}$$

and the prescribed temperatures and their rates

$$\overline{\boldsymbol{T}} = [\overline{T}_j], \quad \dot{\overline{\boldsymbol{T}}} = \left[\frac{\partial \overline{T}_i}{\partial t}\right], \quad \text{prescribed } j .\tag{8.7}$$

Therefore, Eq. (6.34) may be recast as

$$\boldsymbol{C}\dot{\boldsymbol{T}} + \widetilde{\boldsymbol{K}}\boldsymbol{T} + \overline{\boldsymbol{C}\dot{\boldsymbol{T}}} + \overline{\boldsymbol{K}\boldsymbol{T}} - \boldsymbol{L} = \boldsymbol{0} .\tag{8.8}$$

The generalized trapezoidal method proposes to express the relationship between the temperatures and the rates of temperatures at two different time instants, t_n and t_{n+1}, as

$$\theta \dot{\boldsymbol{T}}_{n+1} + (1 - \theta)\dot{\boldsymbol{T}}_n = \frac{\boldsymbol{T}_{n+1} - \boldsymbol{T}_n}{\Delta t} ,\tag{8.9}$$

where a quantity expressed at time t_n is given a subscript n, and $\Delta t = t_{n+1} - t_n$. The free parameter θ is used to control accuracy and stability of the scheme; see Figure 8.1.

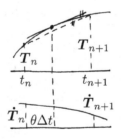

Fig. 8.1   Illustration of the formula (8.9).

Equation (8.9) is applied to the time stepping of (8.8) by writing it at the two time instants, t_n and t_{n+1}, and then mixing together these two equations. Thus, we add together

$$\theta \left[C\dot{T}_{n+1} + \widetilde{K}T_{n+1} + \overline{C}\dot{T}_{n+1} + \overline{K}T_{n+1} - L_{n+1} \right] = 0 , \qquad (8.10)$$

and

$$(1 - \theta) \left[C\dot{T}_n + \widetilde{K}T_n + \overline{C}\dot{T}_n + \overline{K}T_n - L_n \right] = 0 , \qquad (8.11)$$

and if we assume that Eq. (8.9) applies not only to the free temperatures, but also to the prescribed temperatures, the mixture of rates (left-hand side of (8.9)) may be replaced with the divided difference of the temperatures (right-hand side of (8.9)). The resulting equation refers only to temperatures at two time instants, and may be solved to yield T_{n+1}, provided T_n is known.

$$\left[\frac{1}{\Delta t}C + \theta\widetilde{K} \right] T_{n+1} = \left[\frac{1}{\Delta t}C - (1 - \theta)\widetilde{K} \right] T_n + \theta L_{n+1} + (1 - \theta)L_n$$
$$- \left[\frac{1}{\Delta t}\overline{C} + \theta\overline{K} \right] \overline{T}_{n+1} + \left[\frac{1}{\Delta t}\overline{C} - (1 - \theta)\overline{K} \right] \overline{T}_n$$
$$(8.12)$$

The form of Eq. (8.12) is pleasingly symmetric, fully reflective of the blocked nature of these equations. However, for implementation the following form is preferred in SOFEA:

$$\left[\frac{1}{\Delta t}C + \theta\widetilde{K} \right] T_{n+1} = \left[\frac{1}{\Delta t}C - (1 - \theta)\widetilde{K} \right] T_n + \theta L_{n+1} + (1 - \theta)L_n$$
$$- \overline{C}\frac{\overline{T}_{n+1} - \overline{T}_n}{\Delta t} - \overline{K} \left[\theta\overline{T}_{n+1} + (1 - \theta)\overline{T}_n \right]$$
$$(8.13)$$

The last line in this equation indicates how the contributions from prescribed temperatures (and hence also prescribed temperature rates) may be calculated: the term

$$- \overline{C}\frac{\overline{T}_{n+1} - \overline{T}_n}{\Delta t}$$

introduces the contributions of the temperature rates, since the fraction on the right is an approximation of the temperature rate, and the term

$$- \overline{K} \left[\theta\overline{T}_{n+1} + (1 - \theta)\overline{T}_n \right]$$

contributes the effect of prescribed temperatures (in the form of a mixture of temperatures at time n and $n + 1$).

Now to the question of how to choose the value of θ: upon closer inspection of Eq. (8.9) we may conclude that the two choices, $\theta = 0$ and

$\theta = 1$, will lead to Euler methods – the **forward** (explicit) **Euler** for the former, and the **backward** (implicit) **Euler** for the latter. The value of $\theta = 1/2$ is known as the **Crank-Nicholson** method. Figure 8.1 provides a justification for this choice, since if the temperature varied in a parabolic arc between the times n and $n + 1$, the temperature rate at the midpoint would exactly match the slope of the secant.

The explicit Euler method has the limitation of conditional stability, which leads to severe restrictions on the time step. On the other hand, the backward Euler and the Crank-Nicholson are for Eqs. (8.13) unconditionally stable. While the Crank-Nicholson is nominally more accurate than the backward Euler, the latter is often given preference because it tends to eliminate oscillations in the solution.

8.2   The T3 NAFEMS Benchmark

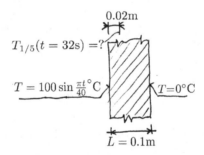

Fig. 8.2   Heat diffusion through a plate (one-dimensional problem), with time-dependent boundary conditions.

This test is recommended by the National Agency for Finite Element Methods and Standards (UK), and it is surprisingly exacting. The domain shown in Fig. 8.2. One face is held at $0°C$, the other face experiences sinusoidal variations in temperature. The temperature at $t = 32$ seconds 0.02 m under the heated face is sought. It is assumed that the plate is very large compared to its thickness, and the problem may therefore be reduced to one dimension, along the thickness. The implementation of transient heat conduction in SOFEA is in fact dimension independent, and we simply take care to define the various objects properly for a 1-D problem and the rest follows.

The solution is presented in the Matlab script **t3nafems**[1]. First, the various parameters are defined. Note that the backward Euler method ($\theta = 1$) is selected for the time discretization.

```
0001 kappa=[35.0]; % conductivity matrix
0002 cm = 440.5;% specific heat per unit mass
0003 rho=7200;% mass density
0004 cv =cm* rho;% specific heat per unit volume
0005 Q=0; % uniform heat source
0006 Tampl=100;
0007 Tamb=0;
0008 Tbar =@(t)(Tampl*sin(pi*t/40)+ Tamb);%hot face temp.
0009 num_integ_pts=2; % quadrature
0010 L=0.1;% thickness
0011 dt=0.5; % time step
0012 tend= 32; % length of the time interval
0013 t=0;
0014 theta = 1.0; % generalized trapezoidal method
0015 online_graphics= ~true;% plot the solution?
0016 n=3*5;% needs to be multiple of five
```

The mesh is created by **block1d**[2], a simple utility which produces a uniformly spaced mesh on the interval $0 \le x \le L$. Note that not only the essential boundary conditions are applied to the temperature field, but also the initial condition (which happens to be 0°C) on line 0032.

```
0017 [fens,gcells] = block1d(L,n,1.0); % Mesh
0018 prop=property_diffusion(struct('conductivity',kappa,...
0019     'specific_heat',cv,'rho',rho,'source',Q));
0020 mater=mater_diffusion (struct('property',prop));
0021 feb = feblock_diffusion (struct (...
0022     'mater',mater,...
0023     'gcells',gcells,...
0024     'integration_rule',gauss_rule(1,num_integ_pts)));
0025 geom=field(struct('name',['geom'],'dim', 1, 'fens',fens));
0026 tempn = field(struct ('name',['temp'], 'dim', 1,...
0027     'nfens',get(geom,'nfens')));
0028 tempn = set_ebc(tempn, 1, 1, 1, Tbar(t));
```

---

[1] Folder: SOFEA/examples/diffusion
[2] Folder: SOFEA/meshing

```
0029 tempn = set_ebc(tempn, n+1, 1, 1, Tamb);
0030 tempn = apply_ebc (tempn);
0031 tempn = numbereqns (tempn);
0032 tempn = scatter_sysvec(tempn,gather_sysvec(tempn)*0+Tamb);
```

The conductivity and capacity matrix are time independent; we compute them once, and henceforth work only with the arrays Km and Cm.

```
0033 K = start (dense_sysmat, get(tempn, 'neqns'));
0034 K = assemble (K, conductivity(feb, geom, tempn));
0035 Km = get(K,'mat');
0036 C = start (dense_sysmat, get(tempn, 'neqns'));
0037 C = assemble (C, capacity(feb, geom, tempn));
0038 Cm = get(C,'mat');
```

The time stepping begins. First, the temperature boundary conditions are time-dependent, which means they have to be set for each pass through the time loop (i.e. for each time instant).

```
0039 Tfifth = [];
0040 while t<tend+0.1*dt % Time stepping
0041     if online_graphics
...
0046     end
0047     tempn1 = tempn;
0048     tempn1 = set_ebc(tempn1, 1, 1, 1, Tbar(t+dt));
0049     tempn1 = set_ebc(tempn1, n+1, 1, 1, Tamb);
0050     tempn1 = apply_ebc (tempn1);
```

The thermal loads corresponding to nonzero temperatures and temperature rates are applied next. We may compare the fields that are being passed on lines 0052 and 0054 with (8.13) and the discussion below that equation: The Matlab code is a literal transcription of the formulas. Note that we are directly working with objects of the class **field**, using operator overload (adding and multiplying fields).

```
0051     F = start (sysvec, get(tempn, 'neqns'));
0052     F = assemble (F, nz_ebc_loads_conductivity(feb, geom,
0053         theta*tempn1 + (1-theta)*tempn));
0054     F = assemble (F, nz_ebc_loads_capacity(feb, geom, ...
0055         (tempn1-tempn)*(1/dt)));
0056     Tn=gather_sysvec(tempn);
```

```
0057      Tfifth = [Tfifth Tn(n/5+1)];
```

The individual objects in this system of linear equations for Tn1 are again directly recognizable in formula (8.13).

```
0058      Tn1=(1/dt*Cm+theta*Km)\((1/dt*Cm-(1-theta)*Km)*Tn+...
0059         get(F,'vec'));
0060      tempn = scatter_sysvec(tempn1,Tn1);
0061      t=t+dt;
0062 end
```

The results are summarized in Fig. 8.3. The reference solution is 36.6°C at the time $t = 32$ seconds, and the curve shown in the figure has been obtained with 500 elements through the thickness (yielding 36.16°C). The solution with 15 elements is seen to be in considerable error. This is somewhat surprising, but a closer look at the behavior of the solution during the time interval of interest shows significant temperature gradients near the hot surface, which provides an explanation of why it is so expensive to get an accurate solution.

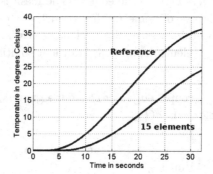

Fig. 8.3   Heat diffusion through a plate (one-dimensional problem): temperature 0.02 m under the heated face.

8.3   Transient cooling in a shrink-fitting application

Shrink fitting is a common manufacturing process used to assemble two parts. The two parts are tempered to different temperatures in order for one to shrink and the other one to expand. Figure 8.4 shows of a cross-section of the three-dimensional steel block into which a tungsten insert is to be fitted. In our case, the cold part is maintained at $-10°C$ prior to the assembly, while the hot part is at $84°C$. The temperature of the ambient air is $17°C$. The task is to determine how long it will take before the temperature of the hot part drops below $75°C$ (which is given as a manufacturing constraint).

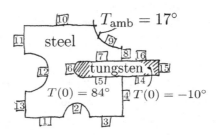

Fig. 8.4   Transient cooling of a shrink-fitted assembly: schematic.

The problem is solved by the script **shrinkfit**[3]. Let us jump directly to the mesh generation: Figure 8.4 shows the two regions, and the boundary edges (note the numbers next to the edges). The mesh is relatively coarse considering the thickness of some of the geometry – only around three elements through the thickness of the cold part. Even so the mesh has almost 2300 triangular elements, and the transient solution takes a couple of minutes.

```
0017 [fens,gcells,groups,edge_gcells,edge_groups]=
            targe2_mesher({...
0018      'curve 1 line 0 0 50 0',...
0019      'curve 2 arc 50 0 80 0 center 65 -0.001 ',...
0020      'curve 3 line 80 0 110 0',...
0021      'curve 4 line 110 0 110 50',...
0022      'curve 5 line 110 50 65 50 ',...
0023      'curve 6 arc 65 50 65 70 center 65.001 60   ',...
0024      'curve 7 line 65 70 110 70',...
0025      'curve 8 line 110 70 110 85',...
0026      'curve 9 arc 110 85 65 120 center 110 120 ',...
0027      'curve 10 line 65 120 0 120',...
0028      'curve 11 line 0 120 0 85',...
0029      'curve 12 arc 0 85 0 35 center -0.001 60 rev',...
```

[3]Folder: SOFEA/examples/diffusion

```
0030      'curve 13 line 0 35 0 0',...
0031      'curve 14 line 110, 50, 160, 50',...
0032      'curve 15 line 160, 50, 160, 70',...
0033      'curve 16 line 160, 70, 110, 70',...
0034      ['subregion 1  property 1 boundary '...
0035      ' 1 2 3 4 5 6 7 8 9 10 11 12 13'],...
0036      ['subregion 2  property 2 boundary '...
0037      ' -5 -6 -7 14 15 16'],...
0038      ['m-ctl-point constant 3']
0039      }, 1.0);
```

The property, material, and block objects are created for each material (steel, tungsten) separately.

```
0040 prop_steel=...
0041  property_diffusion (struct('conductivity',kappa_steel,...
0042  'specific_heat',cv_steel,'source',0.0));
0043 mater_steel=mater_diffusion(struct('property',prop_steel));
0044 feb_steel=feblock_diffusion(struct('mater',mater_steel,...
0045  'gcells',gcells(groups{1}),...
0046  'integration_rule',tri_rule(num_integ_pts)));
0047 prop_tungsten=...
0048  property_diffusion (struct('conductivity',kappa_tungsten,
0049  'specific_heat',cv_tungsten,'source',0.0));
0050 mater_tungsten=mater_diffusion
              (struct('property',prop_tungsten));
0051 feb_tungsten = feblock_diffusion (
              struct ('mater',mater_tungsten,...
0052  'gcells',gcells(groups{2}),...
0053  'integration_rule',tri_rule(num_integ_pts)));
```

For the edges that separate the metal from the air, elements to be used in the surface heat transfer needs to be generated (finite element block **efeb**). Note that the interior edges are omitted.

```
0054 edge_gcells=edge_gcells([edge_groups{[(1:4) (8:16)]}]);
0055 efeb = feblock_diffusion (struct ('mater',mater_steel,...
0056      'gcells',edge_gcells,...
0057      'integration_rule',gauss_rule(1,num_integ_pts),...
0058      'surface_transfer', h));
```

The ambient temperature is defined in the field **amb**. The temperature is applied at the nodes associated with the boundary edges in the loop (line 0063).

```
0062 amb = clone(tempn, ['amb']);
0063 for i= 1:length(edge_gcells)
0064     conn = get(edge_gcells(i),'conn');
0065     amb = set_ebc(amb, conn, conn*0+1, [], conn*0+Ta);
0066 end
0067 amb = apply_ebc (amb);
```

The time stepping loop is almost identical to the one in the example in Section 8.2, except the thermal load vector is based on the convective surface heat transfer. The evolution of the lowest and highest temperature in the entire assembly is shown in Fig. 8.5: the highest temperature drops below 75°C after around 80 seconds. We are not addressing the issue of accuracy, neither in the resolution afforded by the mesh, nor in the selection of the time step. Some pointers to how this could be addressed are given in Chapter 10.

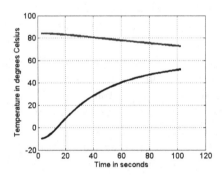

Fig. 8.5   Transient cooling of a shrink-fitted assembly: time evolution of the lowest and highest temperature in the assembly.

Chapter 9

Expanding the Library of Element Types

The linear triangle T3 is not particularly accurate, but for the linear heat conduction problem it is quite adequate. Nevertheless, we will introduce a selection of other elements to expand the scope of the approximation methods discussed so far. This is desirable from a couple of different viewpoints. Firstly, with the linear triangle we have been able to construct basis functions which allow for linear variations in temperature to be represented exactly. Hence, if the exact solution leads to a constant gradient of temperature, the approximate solution does not involve any discretization error. Unfortunately, constant gradients of temperature are not commonly encountered in applications. If the basis functions can represent higher-order polynomials, for instance quadratic, the resulting method will be able to represent more complex gradients of temperature: linear, in the case of the quadratic variation of temperature. Taking the liberty to oversimplify somewhat, *the more complex the temperature variations that are reproduced without error, the higher the overall accuracy of the scheme.*

Secondly, introducing different element types may enable us to play games with different quadrature schemes. One view of the finite element method puts the basis function above all: the elements are there only to integrate all the expressions that involve the basis functions and their derivatives as accurately as possible (exactly?). However, there is a virtue in the complementary view: the basis functions are used essentially to ensure a degree of compatibility (fitting together), and the accuracy of the method can be enhanced by tuning the properties of the individual finite elements. One of the dials that can be used for tuning is the quadrature rule: we shall see how to improve the performance of the elements used for stress analysis by employing integration rules which are on purpose *not* doing their job as accurately as possible.

9.1 Quadratic triangle T6

The triangle T6 makes it possible to design basis functions that can reproduce quadratic variations of the temperature. More precisely, it will do that in terms of the coordinates on the standard triangle. As we shall see, the map from the standard triangle will also allow for quadratic temperature variation in the physical space, but more generally it will lead to rational expressions.

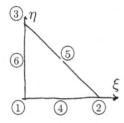

Fig. 9.1   Standard quadratic triangle.

The first task will be to formulate the basis functions on the standard triangle, Fig. 9.1. To be able to write down a polynomial for a particular basis function that is quadratic in ξ, η, six coefficients will be needed. To determine these coefficients, we will make use of the common device of equipping the basis functions with the Kronecker delta property (2.21). Let us start with the basis function $N_2 = a_0 + a_1\xi + a_2\eta + a_3\xi\eta + a_4\xi^2 + a_5\eta^2$. Writing

$$N_2(\xi_k, \eta_k) = \delta_{2k}, \quad \text{for } k = 1, \ldots, 6,$$

at all six nodes (see Table 9.1), provides us with six equations from which the six coefficients may be determined. That is however tedious and boring: let us use commonsense and guesswork instead. Looking along the η axis we see three and two nodes respectively align; see Figure 9.2. Evidently, this makes it possible to design the function N_2 as a Lagrange polynomial that is zero in these two locations, and equal to one at node 2

$$N_2 = \frac{(\xi - 0)(\xi - 1/2)}{(1 - 0)(1 - 1/2)} = \xi(2\xi - 1).$$

Similarly, in the other direction we have for $N_3 = \eta(2\eta - 1)$.

To approach the construction of the other basis functions, we note that both N_2 and N_3 may be written as the normalized product of planes: for

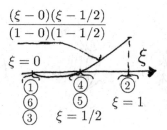

Fig. 9.2 Standard quadratic triangle: one-dimensional view of the basis function N_2.

Table 9.1 Standard quadratic triangle: locations of the nodes

Coordinate	Node 1	Node 2	Node 3	Node 4	Node 5	Node 6
ξ	0	1	0	1/2	1/2	0
η	0	0	1	0	1/2	1/2

N_2 the two planes are $\widehat{p}_2(\xi, \eta) = \xi$ and $\widetilde{p}_2(\xi, \eta) = \xi - 1/2$, and N_2 is written as

$$N_2 = \frac{\widehat{p}_2(\xi, \eta)\widetilde{p}_2(\xi, \eta)}{\widehat{p}_2(1, 0)\widetilde{p}_2(1, 0)} = \xi(2\xi - 1) \ .$$

Similarly for N_3 and N_1: the recipe is to find two planes that go through three nodes and two nodes respectively (but not through the node at which the function is supposed to be equal to one), and normalize their product. For N_1 the planes are $\widehat{p}_1(\xi, \eta) = 1 - \xi - \eta$ (this is the same N_1 as in (6.14)) and $\widetilde{p}_1(\xi, \eta) = 1 - 2\xi - 2\eta$ (compare with Fig. 9.3)

$$N_1 = (1 - \xi - \eta)(1 - 2\xi - 2\eta) \ .$$

For the mid-edge nodes, 4, 5, 6, we find planes that pass through two triples of nodes. For instance, for node 6 (see Fig. 9.3), the two planes are $\widehat{p}_6(\xi, \eta) = 1 - \xi - \eta$ and $\widetilde{p}_6(\xi, \eta) = \eta$ (same as N_3 as in (6.13))

$$N_6 = 4(1 - \xi - \eta)\eta \ .$$

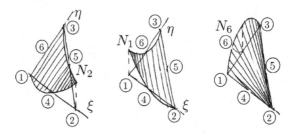

Fig. 9.3   Standard quadratic triangle: Basis functions N_2, N_1, and N_6.

9.2   Quadratic 1-D element L3

The basis functions for this element are simply the Lagrange interpolation functions on the standard interval. In fact, along each edge of the quadratic triangle T6 we have a set of functions which look quite like what we want already. However, the parameters on the standard triangle vary between zero and one; the standard interval is different. Therefore, the basis functions N_1, N_2, and N_3 on the standard interval will read (Fig. 9.4)

$$N_1(\xi) = \frac{\xi(\xi - 1)}{2} \ , \quad N_2(\xi) = \frac{\xi(\xi + 1)}{2} \ , \quad N_3(\xi) = (1 - \xi^2) \ . \tag{9.1}$$

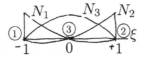

Fig. 9.4   Quadratic basis functions on the standard interval.

9.3   Point element P1

Browsing the `classes/gcell` folder, one may notice the `gcell_X_manifold` class folders with `X=` 0, 1, 2, 3. All SOFEA geometric cells are of certain so-called *manifold dimension*: solids are of manifold dimension 3, surfaces are of dimension 2, while curves and points are of dimensions 1 and 0. Since we commonly solve heat diffusion (and other problems) with functions that are defined in 3-D domains (solids), 2-D domains (surfaces), and 1-D domains (curves), we also have to deal with integration over the boundaries of these domains; these are, correspondingly, surfaces, curves, and points.

When the heat diffusion model was formulated in two-dimensional domains in Chapter 6, the discrete domain consisted of triangles (elements T3), and the discrete boundary consisted of line segments (elements L2). Analogously, when the heat diffusion is solved in a one-dimensional domain (interval of the real line) which is covered by elements L2, the boundary consists of two points: hence the need for elements of type P1.

Evaluating the integrals of the surface heat transfer matrix (6.29) and the surface heat transfer load (6.28) (and also the prescribed heat flux load (6.27)) over the boundary of an interval on the real line simply means taking the values of the integrands at the end points. In terms of a quadrature formula applied at the boundary point a (analogous to (2.26)),

$$f(a) \approx f(\xi_1)J(\xi_1)W_1 \ ,$$

which is going to give the expected results with $\xi_1 = 0$, $f(\xi_1) = f(a)$, $W_1 = 1$, and the Jacobian $J(\xi_1) = 1$. The quadrature rule with these properties is the `point_rule`, and a sample script to use this type of evaluation of boundary integrals for one-dimensional heat diffusion problems is `transcool`[1].

With the introduction of the element P1, a closure is achieved: the same code will now work for heat diffusion problems solved on one-dimensional, two-dimensional, and three-dimensional domains.

9.4 Integrating over n-dimensional domains

The uniform treatment of the manifold dimension of the domain allows us to produce dimension-independent code. Therefore, integration of any (typically, scalar) function over any domain or subdomain is carried out by a single method of the class `feblock`. Consider as an example the geometry of a cylinder, the volume tiled with tetrahedra, the bounding surface covered with triangles, the edges of the cylindrical faces approximated with straight two-node segments, and one node at each vertex of the mesh. We may integrate over the volume of the 3-D mesh to find an approximation of the volume of the original cylinder; or over the length of a single edge to approximate the circumference; or over the area of one circular face to find an approximation of the cross-sectional area; or to count all the nodes on the cylindrical surface when we integrate over all the vertices of the triangles on that surface.

---

[1] Folder: `SOFEA/examples/diffusion`

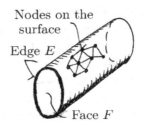

Fig. 9.5   Geometry of a cylinder.

The method `measure` of the class `feblock` evaluates the integral

$$\int_{V_n} f(\boldsymbol{x}) \, dV_n \,, \tag{9.2}$$

where V_n is the volume of an n-dimensional manifold ($n = 0, 1, 2, 3$). There are a number of uses to which the method could be applied: as an example consider the calculation of the moments of inertia, or calculation of the centroid.

The method takes as arguments the geometry field (evidently, the volume of any discrete manifold is going to depend on the locations of its vertices), and a function handle.

```
0014 function result = measure²  (self, geom, fh, varargin)
0015      gcells =self.gcells;
0016      ngcells = length(self.gcells);
0017      % Integration rule
0018      integration_rule = get(self, 'integration_rule');
0019      pc = get(integration_rule, 'param_coords');
0020      w  = get(integration_rule, 'weights');
0021      npts_per_gcell = get(integration_rule, 'npts');
```

The simplest function to be supplied is $f(\boldsymbol{x}) = +1$, which yields as the result of the integration the volume of the manifold. Otherwise, the function can supply for instance the location-dependent mass density. The dimension of the manifold may be deduced from the manifold dimension of the geometric cells, or it could be supplied in the optional `varargin` argument.

```
0022      if nargin >=3
0023          m =varargin{1};
```

---

[2]Folder: `SOFEA/classes/feblock/@feblock`

```
0024     else
0025        m= get(gcells(1),'dim');
0026     end
```

Evaluate the integral by looping over all geometric cells. First step: collect the geometry from the supplied field.

```
0027     result = 0;
0028     % Now loop over all gcells in the block
0029     for i=1:ngcells
0030        conn = get(gcells(i), 'conn'); % connectivity
0031        x = gather(geom, conn, 'values', 'noreshape');
```

Each type of a geometric cell must provide functions for calculating the basis functions and the derivatives of the basis functions with respect to the parametric coordinates in order to evaluate the Jacobian. Again, the method Jacobian_mdim is dispatched dynamically, to be treated differently in dependence on the dimension of the manifold. The result of Jacobian may depend on the location of the point in the parametric or spatial coordinates.

```
0032        % Loop over all integration points
0033        for j=1:npts_per_gcell
0034           detJ = Jacobian_mdim(gcells(i),pc(j,:),x,m);
```

The array of basis functions is computed so that the spatial location may be evaluated, and supplied to the function fh. The argument N'*x is the spatial location of the integration point (it is the interpolation of the locations of the nodes!). The result is accumulated with numerical quadrature.

```
0035           N = bfun(gcells(i), pc(j,:));
0036           result = result + fh(N'*x)*detJ*w(j);
0037        end
0038     end
0039 end
```

As an example, here is the Jacobian_mdim method for a two-dimensional manifold (a surface). It may be a real surface, or it may be equipped with a thickness (or, perhaps, it could be revolved around an axis) and therefore representative of a 3-D volume. The argument m is used to distinguish between these possibilities.

```
0017 function detJ = Jacobian_mdim³(self, pc, x, m)
0018     switch (m)
0019         case 3
0020             detJ = Jacobian_volume(self, pc, x);
0021         case 2
0022             detJ = Jacobian_surface(self, pc, x);
0023         otherwise
0024             error('Wrong dimension');
0025     end
0026 end
```

Let us assume for the moment it is a surface without a thickness, in which case it invokes the method Jacobian_surface to do the work. The number of space dimensions sdim of the space in which the manifold is embedded could be 2 (the manifold is just a piece of the Euclidean plane), or 3 (the manifold is then a piece of a 3-D surface). The number of tangents must be 2 (compare with (6.46), and refer to Fig. 9.6): they are

$$\frac{\partial \boldsymbol{x}(\xi, \eta)}{\partial \xi}, \quad \text{and} \quad \frac{\partial \boldsymbol{x}(\xi, \eta)}{\partial \eta} \ .$$

The Jacobian is the length of the cross product of the two tangents (refer to Fig. 6.9). Here, the cross product is expressed through a skew-symmetric matrix.

```
0013 function detJ = Jacobian_surface⁴(self, pc, x)
0014     Nder = bfundpar (self, pc);
0015     tangents =x'*Nder;
0016     [sdim, ntan] = size(tangents);
0017     if    ntan==2 % 2-D gcell
0018       if sdim==ntan
0019         detJ=det(tangents);% Compute the Jacobian
0020       else
0021         detJ=norm(skewmat(tangents(:,1))*tangents(:,2));
0022       end
0023     else
0024         error('Got an incorrect size of tangents');
0025     end
0026 end
```

---

[3] Folder: SOFEA/classes/gcell/@gcell_2_manifold
[4] Folder: SOFEA/classes/gcell/@gcell_2_manifold

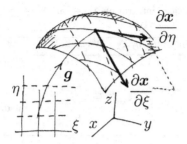

Fig. 9.6   Surface with the coordinate curves and tangents.

Finally, we give an example of the use of the **measure** method: The Matlab script **test_measure**[5] computes the volume and the surface area of a rectangular block tiled with tetrahedra T4. Note that the boundary of the 3-D domain is extracted with the utility **mesh_bdry**

```
0001 a=2.5*pi;
0002 b=2.95;
0003 c=6.1313;
0004 [fens,gcells] = t4block(a,b,c, 5, 4, 7);
0005 bg=mesh_bdry(gcells);
0006 geom=field(struct('name',['geom'],'dim',3,'fens',fens));
0007 feb = feblock (struct ('mater',[], 'gcells',gcells,...
0008     'integration_rule', tet_rule (1)));
0009 disp([' The volume is = '...
0010     num2str(measure(feb,geom,inline('1'))) ...
0011     ', to be compared with ' num2str(a*b*c)])
0012 feb = feblock (struct ('mater',[], 'gcells',bg,...
0013     'integration_rule', tri_rule (1)));
0014 disp([' The surface is = ' ...
0015     num2str(measure(feb,geom,inline('1'))) ...
0016     ', to be compared with ' num2str(2*(a*b+a*c+b*c))])
```

It might seem tempting to evaluate all the objects used in the computational methods of this book, the conductivity matrix, the mass matrix, the load terms, and so on, with a generalization of the **measure** method. Unfortunately, Matlab passes arguments by value, which means that to accumulate as the result, for instance, a 20000×20000 stiffness matrix

---

[5]Folder: SOFEA/examples/miscellaneous

Table 9.2   Numerical integration rules on the standard tetrahedron; $a = 0.1381966$, $b = 0.5854102$.

Rule	Coordinates ξ_j, η_j, ζ_j	Weights W_j	Integrates exactly
1-point	1/4, 1/4, 1/4	1/6	linear polynomial
4-point	a, a, a	1/24	quadratic polynomial
	b, a, a	1/24	
	a, b, a	1/24	
	a, a, b	1/24	

would be prohibitively expensive and wasteful. (The correct result would be produced, if that's any consolation.)

9.5   Tetrahedron T4

The tetrahedron with four nodes at the corners (element T4) is a straightforward extension of the triangle T3. The standard tetrahedron is shown in Fig. 9.7. The basis functions in the parametric coordinates are designed to be linear functions of ξ, η, ζ, and there are four corners at which to use the Kronecker delta property. It is straightforward to deduce that

$$N_1(\xi,\eta,\zeta) = 1-\xi-\eta-\zeta \, , \ N_2(\xi,\eta,\zeta) = \xi \, , \ N_3(\xi,\eta,\zeta) = \eta \, , \ N_4(\xi,\eta,\zeta) = \zeta \, . \tag{9.3}$$

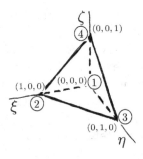

Fig. 9.7   Standard tetrahedron.

Table 9.2 defines two integration rules for tetrahedra [Hughes (2000)]. The one-point rule is adequate for conductivity matrix evaluation, while the four-point rule could handle the capacity matrix terms.

The four basis functions of the tetrahedron each vanish along the opposite face (basis function N_i on the face opposite node i and so on). The remaining three vary along this face exactly as if it was a triangle T3. The situation is entirely analogous to the one discussed in Section 6.9 for the triangle T3 and the line segment L2. Therefore, evaluation of the surface heat transfer contributions for a mesh of T4 volume elements is performed by extracting the faces of the tetrahedra as the geometric cells of type T3, and integrating over those cells.

Example: The script `helixcooled`[6] illustrates a solution with a full 3-D geometry discretized with the T4 tetrahedra. The problem is to determine steady state surface temperature for a helical spring with variable cross-section– see Fig. 9.8. The thick end is maintained at constant temperature, and on the rest of the surface we assume convection cooling.

Fig. 9.8 The cooling of a helical spring.

The mesh is a simple regular block tiled with tetrahedra, but it is then shaped by moving nodes to different locations using the utility `transform_apply`, first by changing its cross-section, and then by shifting all nodes in the y-direction. Finally, the shape is twisted into a helix using `transform_2_helix`.

```
0008 [fens,gcells] = t4block(Angle,Width,Height, 50, 6, 4);
0009 Radius = 1.2;
0010 fens=transform_apply(fens,...
                    @(x,data)(x.*[1,(1-x(1)/Angle/1.2),1]),[]);
0011 fens=transform_apply(fens,@(x,data)(x+ [0,Radius,0]),[]);
0012 climbPerRevolution= 1.3;
0013 fens = transform_2_helix(fens,climbPerRevolution);
```

The surface mesh consists of triangles T3, and is extracted from the tetrahedral mesh using the utility `mesh_bdry`. The surface mesh is drawn with the `drawmesh` utility.

---

[6]Folder: `SOFEA/examples/diffusion`

```
0014 bgcells=mesh_bdry(gcells);
0015 drawmesh({fens,bgcells},'gcells','facecolor','red')
```

Next, the finite element blocks for the tetrahedral elements in the volume
and the triangular elements on the surface are created. Note that the
two blocks use different quadrature rules, tet_rule for the tetrahedra, and
tri_rule for the triangles; both use just one integration point.

```
0017 feb = feblock_diffusion (struct ('mater',mater,...
0018      'gcells',gcells,...
0019      'integration_rule',tet_rule(num_integ_pts)));
0020 bfeb = feblock_diffusion (struct ('mater',mater,...
0021      'gcells',bgcells,...
0022      'integration_rule',tri_rule(num_integ_pts),...
0023      'surface_transfer', h));
```

From this point on, the script does not depend on the element types, be it
the calculation of the system matrices, or graphics output.

9.6   Simplex elements

The point P1, the segment L2, the triangle T3, and the tetrahedron T4,
are all examples of the so-called simplex elements. By definition, an n-
dimensional simplex is the convex hull of $n + 1$ points (vertices) in the
n-dimensional space. Tiling domains with simplex elements is attractive,
because a number of mathematical properties guarantees the success of
automatic tools for mesh generation. This is to be contrasted with the generation of quadrilaterals in two dimensions, and of bricks (shapes bounded
by six quadrilateral faces) in three dimensions: not an easy task– mesh
generators often fail to produce good-quality meshes, or often they just fail
to produce any mesh.

While the simplex elements perform adequately in the heat conduction
models, in other types of analyses their inherent simplicity tends to work
against them. For instance, as we shall see in linear elasticity the response
of meshes composed of simplex elements is quite poorly represented – they
are "too stiff".

9.7   Quadrilateral Q4

Quadrilateral elements address the excessive stiffness of simplex elements by coupling together a larger number of nodes, which in the end leads to basis functions which are more than just linear polynomials.

The element Q4 has four nodes, and its standard shape is a square. This square is to be understood as the Cartesian product of two standard intervals (Fig. 6.12). Therefore, the basis functions of Q4 may also be formed as products of the basis functions on the standard interval L2. Assuming the numbering of the nodes as shown in Fig. 9.9, the basis function N_1 may be written as the product of the basis function on the interval $-1 \leq \xi \leq +1$ and the basis function on the interval $-1 \leq \eta \leq +1$ where both functions correspond to the left-hand side endpoint ($\xi = -1$, $\eta = -1$)

$$N_1(\xi, \eta) = \frac{\xi - 1}{-1 - 1} \times \frac{\eta - 1}{-1 - 1} = \frac{(\xi - 1)(\eta - 1)}{4}. \qquad (9.4)$$

Similarly, for the remaining three functions we have

$$N_2(\xi, \eta) = \frac{(\xi + 1)(\eta - 1)}{-4}, \quad N_3(\xi, \eta) = \frac{(\xi + 1)(\eta + 1)}{4}, \qquad (9.5)$$

and

$$N_4(\xi, \eta) = \frac{(\xi - 1)(\eta + 1)}{-4}. \qquad (9.6)$$

As all basis functions are linear in ξ and η, the shape that they represent when raised as a surface above the standard square is a hyperbolic paraboloid.

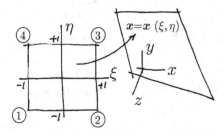

Fig. 9.9   Mapping the standard square to a general quadrilateral.

Since the standard square is a Cartesian product of the standard intervals for which one-dimensional Gauss integration rules are a common choice, a two-dimensional Gauss integration rule is commonly adopted for Q4. It consists of a Cartesian product of one-dimensional Gauss rules. The class **gauss_rule** implements two-dimensional (and three-dimensional) rules which are products of one-dimensional tables. Thanks to the utility **gaussquad** by Peter J. Acklam (included with **SOFEA**), one-dimensional tables of any order may be calculated on demand and used for higher dimensions. For the four-node quadrilateral, a 2 × 2 Gauss quadrature is appropriate for conductivity matrices; a one-point rule is insufficient to build up the proper rank of the element matrices, while higher-order rules are a waste of time. The requirements for the minimum number of integration points from the point of view of stability (regularity of the stiffness matrices) are discussed in Section 16.2.

9.8 Hexahedron H8

To extend the quadrilateral to three dimensions is quite straightforward: instead of a Cartesian product of two intervals on the standard square, we consider the Cartesian product of three intervals on the standard cube (Fig. 9.10). This element is discussed in more detail later, in Section 13.5.

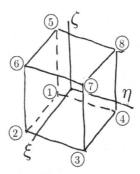

Fig. 9.10   Numbering of the nodes of the hexahedron H8.

9.9 Extracting the mesh boundary

The shapes of the geometric cells in the **SOFEA** toolbox are linked together through the "taking the boundary" operation (symbol ∂ in Fig. 9.11). This capability is crucial because some operations need to be performed over the volume of the mesh, while others should be evaluated over the surface. Also, the volume and surface integrals may need to be computed for models of different number of space dimensions. The general utility `mesh_bdry` may be used to extract the boundary from a mesh or a mesh subset. To support these operations, the geometric cells have the responsibility of computing the connectivity of their boundary and supplying the handle of the constructor of the appropriate boundary geometric cell with their `get` method.

Figure 9.11 summarizes how the various types of geometric cells fit together. Some of these types have been discussed already, some make their appearance later in the book.

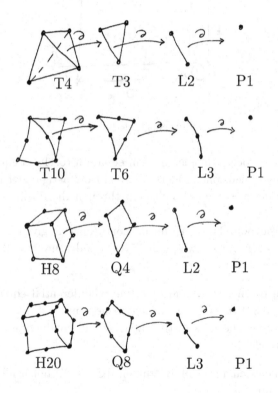

Fig. 9.11   Extracting the boundary from the geometric cells.

Exercises

(1) Compute the conductivity matrix of a single Q4 finite element mesh of Fig. 9.12. The thermal conductivity coefficient of the material is $\kappa = 600 \text{ W/m/K}^\circ$. Evaluate the integrals with 2×2 Gauss quadrature.

    (a) Calculate the rank of the conductivity matrix K, and explain the result.

    (b) Calculate the eigenvectors of K: explain how the individual eigenvectors contribute to the nodal heat power loads.

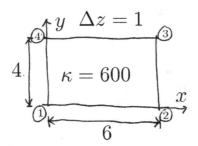

Fig. 9.12   Single-element mesh.

(2) Repeat the process that leads from equation (6.11) to equation (6.30) for the one-dimensional Galerkin model from assignment (Ch. 6-5) for the following mesh: one L3 element through the thickness of the wall (Fig. 9.13).

Assume the boundary and initial conditions as given in the Matlab script **t3nafems**. Use two-point Gauss quadrature for all integrations in the space direction.

    • Write down the system of ordinary differential equations in the form (6.34) specialized to the one-element mesh.

    • Determine the rank of the capacity and conductivity matrices for this model.

(3) Repeat the calculations of assignment (Ch. 6-2), but using a one-point Gauss quadrature.

    • Write down the system of ordinary differential equations in the form (6.34) specialized to the one-element mesh.

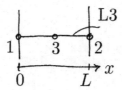

Fig. 9.13 Mesh for the one-dimensional heat conduction problem.

- Determine the rank of the capacity and conductivity matrices for this model.

Chapter 10

Discretization Error, Error Control, and Convergence

In this chapter we will address the error of the finite element approximation. In particular, we will inspect the so-called discretization error, which is the part of the error that is due to the introduction of the finite element basis functions. Also, we will discuss ways of controlling the error.

We begin by outlining how to estimate interpolation errors. The finite element solution in general does not interpolate the exact solution, but it turns out that the interpolation errors are related to the actual errors in the numerical solution. Even though we will not address this relationship, it will prove beneficial to understand the behavior of the different types of errors on the simpler case of the interpolation errors.

10.1 Interpolation errors

We will estimate the difference between the "exact" distribution of temperature, $T(\boldsymbol{x})$, and an interpolation of this function on a finite element mesh, $\Pi_h T(\boldsymbol{x})$. Here h means the mesh "size", or characteristic dimension. Typically, mesh size is taken to mean edge length, or the diameter of the smallest ball that completely encloses an element.

10.1.1 *Interpolation error for temperature*

The interpolating function is defined as

$$\Pi_h T(\boldsymbol{x}) = \sum_k N_k(\boldsymbol{x}) T(\boldsymbol{x}_k) \,, \qquad (10.1)$$

where $\boldsymbol{x}_k$ is the location of the node k, and $T(\boldsymbol{x}_k)$ is the value of the temperature at the location of the node k. For interpolation on a mesh consisting of three-node triangles, when $\boldsymbol{x}$ is in the interior of the element Δ_e, only three basis functions N_k are nonzero at $\boldsymbol{x}$. The basic tool is the

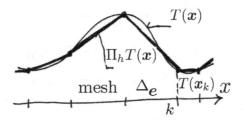

Fig. 10.1   Interpolating the temperature function on a mesh.

Taylor series which we use to expand the temperature at $\boldsymbol{x}$

$$T(\boldsymbol{y}) = T(\boldsymbol{x}) + \text{grad}T(\boldsymbol{x}) \cdot (\boldsymbol{y} - \boldsymbol{x}) + R_1(\boldsymbol{y}, \boldsymbol{x}) \,, \qquad (10.2)$$

where the remainder is written as

$$R_1(\boldsymbol{y}, \boldsymbol{x}) = \frac{1}{2}(\boldsymbol{y} - \boldsymbol{x}) \cdot \boldsymbol{H}(T)(\boldsymbol{y} - \boldsymbol{x}) \,. \qquad (10.3)$$

The matrix of second derivatives (Hessian) is evaluated at $\boldsymbol{\xi}$ somewhere between the points $\boldsymbol{y}$ and $\boldsymbol{x}$

$$[\boldsymbol{H}(T)] = \begin{bmatrix} \dfrac{\partial^2 T(\boldsymbol{\xi})}{\partial x_1 \partial x_1} & \dfrac{\partial^2 T(\boldsymbol{\xi})}{\partial x_1 \partial x_2} \\[2ex] \dfrac{\partial^2 T(\boldsymbol{\xi})}{\partial x_2 \partial x_1} & \dfrac{\partial^2 T(\boldsymbol{\xi})}{\partial x_2 \partial x_2} \end{bmatrix} .$$

The Taylor series (10.2) may be used to express the value of the temperature at the nodes– plug in $\boldsymbol{x}_k$ for $\boldsymbol{y}$– which then may be substituted into the interpolation (10.1) to yield

$$\Pi_h T(\boldsymbol{x}) = \sum_k N_k(\boldsymbol{x})T(\boldsymbol{x}_k) =$$

$$\sum_k N_k(\boldsymbol{x}) \left[T(\boldsymbol{x}) + \text{grad}T(\boldsymbol{x}) \cdot (\boldsymbol{x}_k - \boldsymbol{x}) + R_1(\boldsymbol{x}_k, \boldsymbol{x}) \right] \,.$$

Due to the construction of the basis functions, we have these important equalities

$$\sum_k N_k(\boldsymbol{x}) = 1 \,, \quad \sum_k N_k(\boldsymbol{x})\boldsymbol{x}_k = \boldsymbol{x} \,. \qquad (10.4)$$

Therefore, the first term in (10.4) simplifies as

$$\sum_k N_k(\boldsymbol{x})T(\boldsymbol{x}) = T(\boldsymbol{x})\sum_k N_k(\boldsymbol{x}) = T(\boldsymbol{x}) \,,$$

and the second will vanish

$$\sum_k N_k(\boldsymbol{x})\mathrm{grad}T(\boldsymbol{x}) \cdot (\boldsymbol{x}_k - \boldsymbol{x}) = \mathrm{grad}T(\boldsymbol{x}) \cdot \sum_k N_k(\boldsymbol{x})(\boldsymbol{x}_k - \boldsymbol{x}) = 0 \,.$$

Substituting into (10.4) gives

$$\Pi_h T(\boldsymbol{x}) = T(\boldsymbol{x}) + \sum_k N_k(\boldsymbol{x})R_1(\boldsymbol{x}_k, \boldsymbol{x}) \,,$$

or, reshuffling to get the error on one side,

$$T(\boldsymbol{x}) - \Pi_h T(\boldsymbol{x}) = -\sum_k N_k(\boldsymbol{x})R_1(\boldsymbol{x}_k, \boldsymbol{x}) \,. \tag{10.5}$$

To estimate the magnitude of the difference, $|T(\boldsymbol{x}) - \Pi_h T(\boldsymbol{x})|$, we compute

$$|\sum_k N_k(\boldsymbol{x})R_1(\boldsymbol{x}_k, \boldsymbol{x})| \le \max|R_1(\boldsymbol{x}_k, \boldsymbol{x})| \, |\sum_k N_k(\boldsymbol{x})| = \max|R_1(\boldsymbol{x}_k, \boldsymbol{x})| \,,$$

and make use of standard norm inequalities

$$|\boldsymbol{v} \cdot \boldsymbol{A}\boldsymbol{v}| \le \|\boldsymbol{v}\|\|\boldsymbol{A}\boldsymbol{v}\| \le \|\boldsymbol{A}\|\|\boldsymbol{v}\|^2 \,.$$

This may be applied to the definition of the remainder (10.3) together with (see Fig. 10.2)

$$\|\boldsymbol{x}_k - \boldsymbol{x}\| \le h \,,$$

and an estimate of the norm of the matrix of second derivatives of the temperature $\boldsymbol{H}(T)$ to give

$$|T(\boldsymbol{x}) - \Pi_h T(\boldsymbol{x})| \le Ch^2\|\boldsymbol{H}(T)\| \,. \tag{10.6}$$

Here C is a generic constant with respect to h. If we wrap the norm of the matrix of the second derivatives into the constant, we may write

$$|T(\boldsymbol{x}) - \Pi_h T(\boldsymbol{x})| \le C\left(\partial^2 T\right)h^2 \,, \tag{10.7}$$

where we agree to mean by $C\left(\partial^2 T\right)$ some constant whose magnitude depends on the curvatures of the function T. Importantly, $C\left(\partial^2 T\right)$ may also be understood as measuring the *rate of change of the heat flux* in the immediate neighborhood of $\boldsymbol{x}$.

Fig. 10.2   Mesh size h as a diameter of an element.

The value of Eq. (10.7) is twofold:

- Firstly, it states that the errors of interpolation will get bigger the higher the curvature of the function of the exact temperature T (that is the faster the heat flux is changing) and the bigger the elements (i.e. the error will increase with h^2);
- Secondly, if we are interested in the interpolation error at a particular location, we may consider the curvatures at that location as given, and the Eq. (10.7) then says that the error will decrease as $O(h^2)$ as $h \to 0$ (order-of estimate: reduce h with a factor of two, and the error will decrease with a factor of four).

10.1.2   *Interpolation error for temperature gradient*

To estimate errors for the gradient of temperature, we start with the interpolation (10.4), of which we take the gradient

$$
\mathrm{grad}\Pi_h T(\boldsymbol{x}) = \sum_k \mathrm{grad} N_k(\boldsymbol{x}) T(\boldsymbol{x}_k) =
$$

$$
\sum_k \mathrm{grad} N_k(\boldsymbol{x}) \left[T(\boldsymbol{x}) + \mathrm{grad}T(\boldsymbol{x}) \cdot (\boldsymbol{x}_k - \boldsymbol{x}) + R_1(\boldsymbol{x}_k, \boldsymbol{x}) \right] =
$$

$$
\sum_k \mathrm{grad} N_k(\boldsymbol{x}) T(\boldsymbol{x}) + \sum_k \mathrm{grad} N_k(\boldsymbol{x}) \mathrm{grad}T(\boldsymbol{x}) \cdot (\boldsymbol{x}_k - \boldsymbol{x})
$$

$$
+ \sum_k \mathrm{grad} N_k(\boldsymbol{x}) R_1(\boldsymbol{x}_k, \boldsymbol{x}) =
$$

$$
T(\boldsymbol{x}) \sum_k \mathrm{grad} N_k(\boldsymbol{x}) + \mathrm{grad}T(\boldsymbol{x}) \cdot \sum_k \mathrm{grad} N_k(\boldsymbol{x})(\boldsymbol{x}_k - \boldsymbol{x})
$$

$$
+ \sum_k \mathrm{grad} N_k(\boldsymbol{x}) R_1(\boldsymbol{x}_k, \boldsymbol{x}) \, .
$$

$$
(10.8)
$$

Differentiating (10.4) we obtain

$$\sum_k \text{grad}N_k(\boldsymbol{x}) = 0 \ , \quad \sum_k \boldsymbol{x}_k \text{grad}N_k(\boldsymbol{x}) = \boldsymbol{1} \ , \tag{10.9}$$

which upon substitution into (10.8) yields

$$\text{grad}\Pi_h T(\boldsymbol{x}) = \text{grad}T(\boldsymbol{x}) + \sum_k \text{grad}N_k(\boldsymbol{x})R_1(\boldsymbol{x}_k, \boldsymbol{x}) \ .$$

$$\tag{10.10}$$

Again, an estimate of the magnitude is desired,

$$|\text{grad}T(\boldsymbol{x}) - \text{grad}\Pi_h T(\boldsymbol{x})| = |\sum_k \text{grad}N_k(\boldsymbol{x})R_1(\boldsymbol{x}_k, \boldsymbol{x})| \leq$$

$$\max|R_1(\boldsymbol{x}_k, \boldsymbol{x})| \sum_k |\text{grad}N_k(\boldsymbol{x})| \ .$$

To estimate the magnitude of the gradient of the basis function, we invoke

Fig. 10.3 Triangle quality measures using the radius of the inscribed circle and the diameter of the circumscribed circle. Good (almost equilateral) triangle on the left; bad triangles (obtuse, needle-like) on the right.

the picture of the basis function as a plane that assumes value one at one node and drops off to zero along the opposite edge. Therefore, the largest magnitude of the basis function gradient will be produced by the smallest height in the triangle. The shortest height $d_{\min}$ may be estimated from the radius of the largest inscribed circle, ρ (see Fig. 10.3), as $d_{\min} \approx O(\rho)$. This can be linked to the so-called "shape quality" of a triangle using the **quality measure**

$$\gamma = \frac{h}{\rho} \ ,$$

as $d_{\min} \approx O(\gamma^{-1})h$. The magnitude of the basis function gradient may be then estimated as

$$\max \mathrm{grad} N_k(\boldsymbol{x}) = \frac{1}{d_{\min}} \approx \frac{\gamma}{h} \; .$$

Putting everything together, we obtain

$$|\mathrm{grad} T(\boldsymbol{x}) - \mathrm{grad} \Pi_h T(\boldsymbol{x})| \leq C h^2 \frac{\gamma}{h} \|\boldsymbol{H}(T)\| = C \left(\partial^2 T \right) \gamma h \; . \quad (10.11)$$

The value of Eq. (10.11) is again twofold:

- Firstly, it states that the errors of interpolation for the gradient of temperature will get bigger the higher the curvature of the function of the exact temperature T, the larger the elements (i.e. the error will increase with h), and the larger the quality measure γ (i.e. the worse the shape of the triangle);
- Secondly, considering the curvatures at a fixed location as given, the equation (10.11) states that the error will decrease as $O(h)$ as $h \to 0$ (note that this is one order lower than for the temperatures themselves: reduce h with a factor of two, and the error will decrease with the same factor).

Importantly, Eq. (10.11) allows us to make a general observation: the quantity calculated in the finite element solution is the temperature, the gradient (or, alternatively, the heat flux) is obtained by *differentiation* of the computed temperature, which immediately results in a reduction of the order of dependence on the mesh size. Phrased differently: the temperature results will converge faster than the temperature-gradient results (or, equivalently, heat flux results), because h^2 approaches zero much quicker than h as $h \to 0$.

10.1.3   *Controlling the error; Convergence rate*

That the error depends on the mesh size is very important. The mesh size is in fact one of the things we can use to control the error. If we set up the finite element procedure to solve the problem repeatedly, changing the mesh size to reflect the distribution of error (large error – small elements), we obtain the so-called **adaptive refinement** technique, or **$h-$adaptive refinement method**. The h stands for the mesh size as the control of the error. On the other hand, we could try to reduce the error by increasing the number of terms matched in the Taylor series (10.2). This could be achieved by using higher order polynomials as basis functions. The resulting procedure would be called the *p−**adaptive refinement method***, where the p stands for the polynomial order as the control of the error.

In this book, we will use an increase in polynomial order of the elements only to increase the approximation capacity of the finite element method in one shot (by selecting element type, linear or quadratic), not adaptively (which would involve increasing the polynomial order in a targeted fashion, locally, and to much higher order than just quadratic).

Therefore, let us discuss the errors and how to control them from the point of view of the h−adaptive refinement method. The error must be dependent on a *positive* (but not necessarily integral) *power* of h. Only then decreasing h will lead to a reduction of the error. For instance, for quantity q we require for the error

$$E_q(h) = q_{\text{ex}} - q_h \approx Ch^\beta , \quad \beta > 0 ,$$

as $h \to 0$. The exponent of the mesh size β is called the ***convergence rate*** (or rate of convergence).

Equation (10.11) tells us how to reduce the error. Consider that at a given location, $C\left(\partial^2 T\right)$ cannot be controlled (it is determined by the behavior of the exact solution). What we can influence is the shape of the elements (γ), and the mesh size (h). Now let the point $\boldsymbol{x}$ range across the computational domain. At some locations $C\left(\partial^2 T\right)$ is small, and at others it is large. As an illustration, let us contemplate Fig. 10.4 (and the close-up in Fig. 10.5). The constant $C\left(\partial^2 T\right)$ will be large where the heat flux changes a lot; therefore, where the red arrows which indicate the magnitude and direction of the heat flux strongly change direction or stretch or shrink, the error constant $C\left(\partial^2 T\right)$ should be expected to be large. Intuitively, those are the locations where reducing the error will make a difference. In these locations, the elements should be made smaller. How much? The answer to that question is somewhat elusive: as much as required to reduce the error below desired level, but the precise size to effect this reduction is not known in general. Automatic procedures to estimate the *relative* desired mesh size are becoming available in commercial softwares. To summarize, the h−adaptive method will attempt to control the error by designing ***graded meshes***, with small elements located in regions of expected high error, and proportionally large elements elsewhere.

A very good indication of large errors are the so-called ***reentrant corners*** (concave corners), where the solution typically displays singularities in the form of infinite curvature(s) of the temperature directly in the

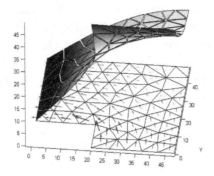

Fig. 10.4   The effect of a reentrant corner on the flux. Matlab script `lshape2`.

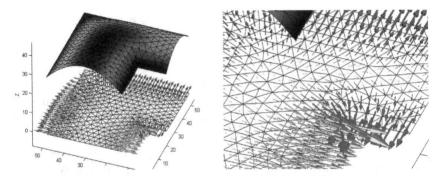

Fig. 10.5   The effect of a reentrant corner on the flux: overall view and close-up. Matlab script `lshape3`.

corner. Figure 10.6 shows an appropriately refined mesh around the reentrant corner for the problem from Fig. 10.5.

10.2   Richardson extrapolation

The second important use of the fact that the error decreases asymptotically as some power of the mesh size is a procedure to improve the estimate of the exact answer based on a series of calculated solutions: in other words, mesh-size-based extrapolation.

Richardson extrapolation is a way of extracting an asymptotic estimate of some quantity of interest from a series of computed values for it. If we assume that the error in the quantity q may be expanded in a Taylor series

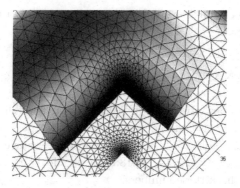

Fig. 10.6 The effect of a reentrant corner on the flux: adaptively refined mesh. Matlab script `1shape3ad`.

at mesh size $h = 0$, we may write

$$E_q(h) = q_{\text{ex}} - q_h \approx C h^\beta, \tag{10.12}$$

where q_{ex} is the unknown true value of the quantity, q_h is the approximate value for nonzero h, C is an unknown constant of the leading term h^β, with β, again, unknown. Provided C, and β do not depend on h for small mesh sizes (this is presumed to hold in the so-called **asymptotic range**), we might be able to compute all three q_{ex}, C, and β, if three numerical solutions are obtained for three different mesh sizes, q_{h_i}. (It does not matter whether the solutions are obtained with uniform or graded meshes.)

Remarkably, the estimate of the exact solution is available from a very easily solvable equation if the condition

$$\frac{h_1}{h_2} = \frac{h_2}{h_3},$$

holds, as we may then combine

$$\frac{q_{\text{ex}} - q_{h_1}}{q_{\text{ex}} - q_{h_2}} = \frac{h_1^\beta}{h_2^\beta} \quad \text{and} \quad \frac{q_{\text{ex}} - q_{h_2}}{q_{\text{ex}} - q_{h_3}} = \frac{h_2^\beta}{h_3^\beta},$$

to yield

$$q_{\text{ex}} = \frac{q_{h_2}^2 - q_{h_1} q_{h_3}}{2 q_{h_2} - q_{h_1} - q_{h_3}}. \tag{10.13}$$

It is then straightforward to extricate the other two quantities. The constant C is of limited value, but the exponent β is the **rate of convergence**.

The computation is implemented in the toolbox SOFEA in the utility function richextrapol.

10.3   The T4 NAFEMS Benchmark revisited

This problem has been discussed in Section 7.4. The publication [Cameron et al. (1994)] cites the reference value for the temperature at the point indicated in Fig. 7.5 of 18.3°C. However, more recent investigations of this benchmark indicate that value of 18.25°C should be expected [Infolytica (2005)]. Let us check these numbers.

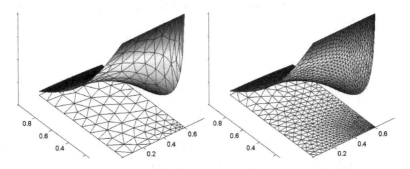

Fig. 10.7   T4 NAFEMS Benchmark: solution with quadratic elements, initial and final mesh.

Two models will be used, the first using elements T3, and the second using the more accurate quadratic elements T6. The Matlab script t4nafems_conv[1] runs the simulation (the initial and final adaptive meshes are shown in Fig. 10.7), with the following results: For the quadratic elements, the Richardson extrapolation produces an estimate of the exact temperature 18.25396°C and the rate of convergence 2.2945. On the other hand, the element T3 performs erratically, and no asymptotic estimate is possible. That is clearly visible in Fig. 10.8: it should be possible to pass a straight line through the estimates of the error if the data is indeed in the asymptotic range, as taking a logarithm of (10.12) yields

$$\log |E_q(h)| = \log |q_{ex} - q_h| \approx \log C + \beta \log h ,$$

which is a straight line on a log-log scale (Fig. 10.8). That is out of the question for the element T3.

---

[1]Folder: SOFEA/examples/diffusion

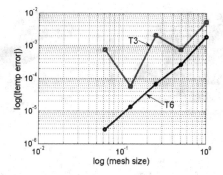

Fig. 10.8   T4 NAFEMS Benchmark: convergence in terms of an estimated error for linear and quadratic triangles.

10.4   Graded meshes

A word on the meaning of the *mesh size in graded meshes* is in order. The mesh size in graded meshes varies from point to point (as opposed to uniform meshes, where the mesh size does not vary). However, if we take one particular graded mesh M_0, and produce a series of meshes from M_0 by scaling the mesh size as a function of x by the same number everywhere in the domain, we may take as the mesh size the scaling factor (the absolute values do not matter, only the relative changes $h_1/h_2 = h_2/h_3$). For instance, mesh M_1 would be produced with mesh size $h_1(x) = \alpha h_0(x)$, mesh M_2 with mesh size $h_2(x) = \alpha h_1(x) = \alpha^2 h_0(x)$, and so on.

10.5   Shrink fitting revisited

Figure 10.9 shows the temperature distribution at three time instants. The extremely high gradient at the beginning is evident, but in fact high temperature gradients exist even at the end of the process.

As you will recall, the heat flux is derived from the temperature (Eq. (5.14)). The finite element approximation with the triangles (T3) and with the line elements (L2) will be able to reproduce linearly varying temperatures, hence constant temperature gradients (i.e. heat flux). Therefore, we will conclude that where the heat flux changes, the finite element approximation will be in error. To control the error, we can reduce the element dimensions. Doing so in areas of steep changes in the heat flux, while keeping areas with approximately uniform heat flux tiled with coarse

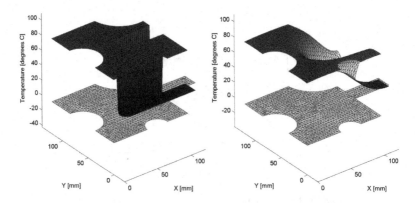

Fig. 10.9   Transient cooling of a shrink-fitted assembly; left to right: temperature distribution for time $t = 0$ and $t = 10$ seconds.

elements, is known as ***adaptive mesh control***.

Figure 10.10 shows the heat flux on two meshes as arrows centered at the barycenters of the elements (barycenter here means average of the vertex locations). The first mesh is quite coarse (script `shrinkfitad1`[2]), but it is possible to identify regions in which the gradient changes strongly (next to the tungsten inset); the graded mesh is generated to reflect the demand for finer (smaller) elements (script `shrinkfitad2`[3]).

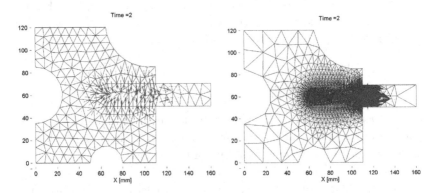

Fig. 10.10   Transient cooling of a shrink-fitted assembly; left: coarse mesh, right: adaptive mesh. Heat flux for time $t = 2$.

The temperature evolution obtained with the two meshes, the coarse one, and the adaptively refined one, is illustrated in Fig. 10.11, and the

---

[2] Folder: `SOFEA/examples/diffusion`
[3] Folder: `SOFEA/examples/diffusion`

higher-quality of the adaptive results should be noted: especially striking is the spurious oscillation of the highest temperature for the coarse uniform mesh.

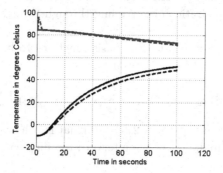

Fig. 10.11 Transient cooling of a shrink-fitted assembly: time evolution of the lowest and highest temperature in the assembly. Comparing temperatures obtained with a coarse model (dashed lines) and with an graded (adaptively refined) model (solid lines).

10.6 Representing functions by interpolation

Using the interpolation (10.1) we can approximate arbitrary functions on finite element meshes. It is of considerable interest to ask which of these functions may be represented exactly (**reproduced**). In other words, the question is for which functions $f(x)$ do we get the same function back,

$$\Pi_h f(x) = f(x) \,,$$

when it is interpolated? To explore the issue, we will pick a quadratic element, the line element L3. The basis functions for this element are given on the standard interval by Eqs. (9.1). The element L3 is an isoparametric element, meaning that the Cartesian coordinates of the nodes are interpolated using (6.16) to yield

$$x = N_1(\xi)x_1 + N_2(\xi)x_2 + N_3(\xi)x_3 \,, \tag{10.14}$$

which is a map of the form of (6.56). For simplicity, we consider the map to send ξ to an interval on the real line. Under suitable conditions, this map may be inverted to yield

$$\xi = \Gamma(x) \,.$$

If we substitute $\Gamma(x)$ for ξ in Eqs. (9.1), do we get basis functions $N_k(x)$ that are quadratic in x? Provided this is the case, we can answer our original question: if we interpolate three (distinct!) data points produced by a quadratic function f using quadratic basis functions, we will get back precisely f: The quadratic curve passing through these three data points is unique.

What are the conditions for the basis functions to be quadratic in x? Expanding the map (10.14), we obtain

$$x = N_1(\xi)x_1 + N_2(\xi)x_2 + N_3(\xi)x_3 = \frac{\xi^2}{2}(x_1 + x_2 - 2x_3) + \frac{\xi}{2}(x_2 - x_1) + x_3 .$$

Now notice that when $(x_1 + x_2 - 2x_3) = 0$ (that is, when $x_3 = (x_1 + x_2)/2$: the node x_3 is the midpoint of the interval), the map will be linear

$$x = \frac{\xi}{2}(x_2 - x_1) + x_3 .$$

For such a linear map, $\Gamma(x)$ is also linear in x, and an expression that is quadratic in ξ (namely the basis functions (9.1)), will be quadratic in x when we substitute for ξ. Therefore, if the node x_3 is the midpoint of the interval $x_1 \leq x \leq x_2$, the complete *quadratic function* $f(x) = ax^2 + bx + c$ will be *reproduced exactly* by the finite element interpolation on the element L3

$$\Pi_h f(x) = \sum_{k=1}^{M} N_k(x)f(x_k) = f(x) .$$

Another way of expressing the restriction on the form of the map is to say that the Jacobian must be constant (and positive).

No such restriction is required if we're interested in reproducing only linear functions $f(x) = bx + c$. The degrees of freedom are set as

$$f_k = bx_k + c ,$$

and we obtain

$$\Pi_h f(x) = \sum_{k=1}^{M} N_k(x)f_k = \sum_{k=1}^{M} N_k(x)\left[bx_k + c\right] .$$

The interpolation of the Cartesian coordinates (6.16) gives

$$\sum_{k=1}^{M} N_k(x)bx_k = b\sum_{k=1}^{M} N_k(x)x_k = bx ,$$

and the partition of unity property (6.15) yields

$$\sum_{k=1}^{M} N_k(x)c = c \sum_{k=1}^{M} N_k(x) = c \,.$$

Therefore, for $f(x)$ linear, we get $\Pi_h f(x) = f(x)$.

These observations may be generalized to all the elements discussed in this book: because all are isoparametric, linear functions may be reproduced exactly by interpolating on the mesh. Furthermore, provided the mapping from the standard shape to the element shape in the physical coordinates has a constant Jacobian, and provided the number of parametric and physical coordinates match, polynomials that are included in the basis functions on the standard shape will be reproduced also in the physical coordinates.

Exercises

(1) Use the Matlab script `circle_area`[4] to compute the approximate area of the circle by integrating over a mesh of triangles.

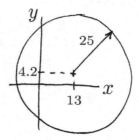

Fig. 10.12   Circle offset with respect to the coordinates xy.

(a) Derive the error as an order-of estimate in terms of the mesh size, $E_A \approx O(h^\beta)$, where h is the mesh size (length of a typical mesh edge).

(b) Relate the error estimate to the experimental data in the graph of the (log of) error versus the (log of) mesh size.

---

[4]Folder: SOFEA/examples/miscellaneous

(2) Modify the Matlab script `circle_area`[5] from assignment (1) to compute the location of the centroid of the circle by integrating over a mesh of triangles.

(3) Modify the Matlab script `circle_area`[6] to compute all the moments of inertia (seconds moments) of the circle by integrating over a mesh of triangles.

    (a) Derive the error as an order-of estimate in terms of the mesh size, $E_A \approx O(h^\beta)$, where h is the mesh size (length of a typical mesh edge).

    (b) Relate the error estimate to the experimental data in the graph of the (log of) error versus the (log of) mesh size.

(4) Repeat the process of assignment (3), but using a three-point quadrature (`tri_rule(3)`).

    (a) Derive the error as an order-of estimate in terms of the mesh size, $E_A \approx O(h^\beta)$, where h is the mesh size (length of a typical mesh edge).

    (b) Relate the error estimate to the experimental data in the graph of the (log of) error versus the (log of) mesh size.

    (c) Explain the difference between the results obtained with the two different quadrature rules.

---

[5] Folder: `SOFEA/examples/miscellaneous`
[6] Folder: `SOFEA/examples/miscellaneous`

Chapter 11

Model of Elastodynamics

We can consider a deformable body to be a collection of particles, and apply the Newton's equation of motion of elementary dynamics to each particle, $m\dot{v} = F$, where $\dot{v}$ is the particle acceleration, m is the particle mass, and F is the applied force. The complicating circumstance is that a deformable body can be thought of as a collection (of infinitely many) particles, all interacting through contact. Furthermore, our goal is to formulate a continuum model rather than deal with the discrete collection of particles.

11.1 Balance of linear momentum

Let us consider a body with some distributed force on parts of the boundary (the reactions must be included) and distributed force in the volume (for instance, gravity-induced load). For simplicity, we draw a sketch in two dimensions, but obviously we are thinking of a three-dimensional body; see Fig. 11.1. The distributed force on the boundary is therefore in units force/length^2, and units of the distributed force in the volume are force/length^3. The distributed force on the boundary is customarily called the *traction*.

The continuous body will be now divided into many very small (infinitesimally small) volumes, which we may consider "particles". The interaction between the particles is mediated by contact forces (tractions) along the cuts between the particles. Assuming we know these forces, the Newton's equation may be applied to each separately. However, we will apply this equation in the form of the change of *linear momentum*

$$\frac{\mathrm{d}}{\mathrm{d}t}(mv) = F ,$$

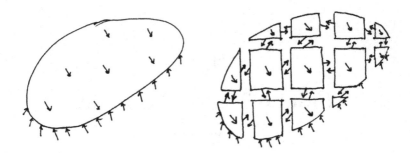

Fig. 11.1   A continuous body with applied distributed force on the boundary, and within the volume (on the left). The same body cut up into many small volumes (particles), with their interaction represented by distributed forces along the cuts (on the right).

from which the previous form of the equation of motion may be obtained provided m does not change. In our case, this will be true because each small volume holds a certain amount of material and does not exchange material with any other volume, so the mass of each volume is conserved.

As a consequence of the above, we may write for each small particle volume j the change of its linear momentum

$$\frac{\mathrm{d}}{\mathrm{d}t}(m_j \boldsymbol{v}_j) = \boldsymbol{F}_j \ , \tag{11.1}$$

where we use for the mass of the particle $m_j = \rho V_j$, with V_j the volume of the particle, and ρ the mass density, $\boldsymbol{v}_j$ the velocity, all at some point within the volume of the particle (we are using the mean-value theorem to express integrals over the volume of the particle!). The force $\boldsymbol{F}_j$ includes the body force $\overline{\boldsymbol{b}}$ and the tractions $\boldsymbol{t}$ on the surface of the particle volume

$$\boldsymbol{F}_j = \overline{\boldsymbol{b}}V_j + \int_{S_{\text{int}}} \boldsymbol{t}\mathrm{d}S + \int_{S_{\text{ext}}} \boldsymbol{t}\mathrm{d}S \ , \tag{11.2}$$

where the surface integral is split into two parts (see Fig. 11.2): the interior surfaces S_{int}, where two particle volumes are separated, and the exterior surfaces S_{ext}.

Now we will collect the contributions of Eq. (11.1) by summing over all the particles

$$\sum_{j=1}^{N} \frac{\mathrm{d}}{\mathrm{d}t}(m_j \boldsymbol{v}_j) = \sum_{j=1}^{N} \boldsymbol{F}_j \ , \tag{11.3}$$

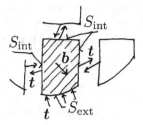

Fig. 11.2 Isolated particle volume.

which may be rewritten in the limit of infinitely many particles as integrals

$$\frac{\mathrm{d}}{\mathrm{d}t} \int_m v \mathrm{d}m = \int_V b \mathrm{d}V + \int_{S_{\mathrm{ext}}} t \mathrm{d}S + \sum_{j=1}^{\infty} \int_{S_{\mathrm{int},j}} t \mathrm{d}S , \qquad (11.4)$$

where the last term (the sum) is over all the shared surfaces that separate the particle volumes. Using Newton's third law of action and reaction, we may conclude that whenever two particle volumes share a piece of their boundary, the traction at the material point A on the surface of particle 1 is equal in magnitude but opposite to the traction at the same material point (the one that has been split by the cut separating the two particles) at the corresponding point A on the surface of particle 2. Since the sum is over all the *pairs* of such surfaces, the last term in Eq. (11.4) cancels, and the final statement of the **balance of linear momentum** of the material in the volume V reads

$$\frac{\mathrm{d}}{\mathrm{d}t} \int_m v \, \mathrm{d}m = \int_V b \, \mathrm{d}V + \int_S t \, \mathrm{d}S , \qquad (11.5)$$

where m is the total mass of the material inside the volume V, and S is the bounding surface of the volume V. While the surface S and the volume V change with deformation, and hence are time-dependent, the total mass of the material m does not change (the same particles that were inside the volume before deformation are there during the deformation).

11.2 Stress

The traction vector t may be written in terms of components in a surface-aligned Cartesian basis as $t = t_n n + t_1 e_1 + t_2 e_2$, where t_n is the normal component, and t_k are the shear components. The Cartesian basis is defined at the given point on the surface by first taking the (outer, unit) surface normal as the third basis vector, and then picking arbitrary orthogonal directions in the tangent plane– see Fig. 11.3. The normal component is extracted as

$$t_n = \boldsymbol{n} \cdot \boldsymbol{t} \ . \tag{11.6}$$

The shear part of the traction $\boldsymbol{t}_s$ is obtained by subtracting the normal part of the traction from the traction vector $\boldsymbol{t}$

$$\boldsymbol{t}_s = \boldsymbol{t} - t_n \boldsymbol{n} \ . \tag{11.7}$$

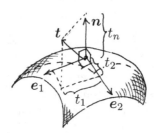

Fig. 11.3   Components of traction.

Next we need to relate the traction on the surface to the deformation of the material just below the surface. The deformation will be measured by strains, and the response of the material to the strains will be related to the tractions on the surface (and any body loads, if present) through the mathematical device of the **stress**.

First, inspect Fig. 11.4: it is possible to define such a Cartesian coordinate system in the vicinity of a given point that the coordinate planes will cut out a (curvilinear) tetrahedron from the solid. Our plan is to make this tetrahedron very small indeed, but to still contain the given point on the surface. An enlarged image of such a tetrahedron is shown on the right, and we see how the curved edges may be approximated by straight lines in the limit of a very small tetrahedron. The goal is to relate the traction at the given point to the tractions on the internal cut planes, because these tractions are representations of the stress in the volume.

In anticipation of the definition of stress, the traction components on the three flat cut planes, with normals pointing against the three Cartesian basis vectors, are called σ_x, σ_y, σ_z (the normal components), and τ_{xy}, τ_{yx}, τ_{xz}, τ_{zx}, τ_{yz}, τ_{zy}, for the shear components on all three planes. The areas

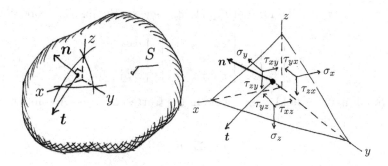

Fig. 11.4   Relating the components of traction to stress.

of the triangular faces of the tetrahedron are related as $A_x = n_x A$, and so forth, where n_x, n_y, n_z are the components of the unit normal, and A_x is the area perpendicular to the x-axis and so on; this can be deduced from the volume of the tetrahedron in Fig. 11.5 written in terms of the heights d, i_x, i_y, i_z, and the corresponding areas.

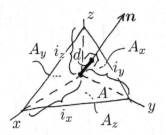

Fig. 11.5   Components of traction.

When we write the conditions of equilibrium in all three directions (the volume forces do not play a role; why?), the following three equations result

$$t_x = \sigma_x n_x + \tau_{xy} n_y + \tau_{xz} n_z \,,$$
$$t_y = \tau_{yx} n_x + \sigma_y n_y + \tau_{yz} n_z \,, \qquad (11.8)$$
$$t_z = \tau_{zx} n_x + \tau_{zy} n_y + \sigma_z n_z \,.$$

This equation relates the components of the traction on the surface with the components of the traction on the special surfaces – coordinate planes – inside the volume. The components of the traction on the internal surfaces are called **normal stresses** (σ_x, σ_y, σ_z), and **shear stresses** (τ_{xy}, τ_{yx},

τ_{xz}, τ_{zx}, τ_{yz}, τ_{zy}). The form of Eq. (11.8) suggests the matrix expression

$$\begin{bmatrix} t_x \\ t_y \\ t_z \end{bmatrix} = \begin{bmatrix} \sigma_x & \tau_{xy} & \tau_{xz} \\ \tau_{yx} & \sigma_y & \tau_{yz} \\ \tau_{zx} & \tau_{zy} & \sigma_z \end{bmatrix} \begin{bmatrix} n_x \\ n_y \\ n_z \end{bmatrix} , \tag{11.9}$$

where all matrices hold components in the Cartesian basis. A component-free version would read

$$ t = \Sigma \cdot n , $$

where Σ would be defined as a Cartesian tensor, the **Cauchy stress tensor**. The traction vector and the normal would then also become tensors. However, in this book the tensor notation is avoided, and with a few exceptions tensors will not be needed. The two exceptions that may be mentioned here are coordinate transformations and the calculation of the **principal stresses** which are the eigenvalues of the matrix of the stress components. The principal direction components and the principle stress σ are solved for from the two equations

$$\begin{bmatrix} \sigma_x & \tau_{xy} & \tau_{xz} \\ \tau_{yx} & \sigma_y & \tau_{yz} \\ \tau_{zx} & \tau_{zy} & \sigma_z \end{bmatrix} \begin{bmatrix} n_x \\ n_y \\ n_z \end{bmatrix} = \sigma \begin{bmatrix} n_x \\ n_y \\ n_z \end{bmatrix} , \tag{11.10}$$

and

$$\det \left(\begin{bmatrix} \sigma_x & \tau_{xy} & \tau_{xz} \\ \tau_{yx} & \sigma_y & \tau_{yz} \\ \tau_{zx} & \tau_{zy} & \sigma_z \end{bmatrix} - \sigma \begin{bmatrix} 1 & 0 & 0 \\ 0 & 1 & 0 \\ 0 & 0 & 1 \end{bmatrix} \right) = 0 . \tag{11.11}$$

The meaning of Eq. (11.10): the traction on the surface given by the normal has only the normal component, the shear components are zero.

11.2.1 *Balance of angular momentum and stress symmetry*

It would appear that there are nine components of stress that need to be related to the deformation, but it is straightforward to show that in the matrix (11.9) the elements reflected with respect to the diagonal must be equal: Consider a rectangular volume of material (again, for convenience the drawing in Fig. 11.6 is of a two-dimensional nature, but the argument applies to three dimensions). When the **balance of angular momentum** is written for the rotation about the axis perpendicular to the plane of the paper, the normal stresses and any body forces will turn out to be negligible compared to to the effect of the shear stresses, and from the resultant equation we obtain the symmetries

$$\tau_{xy} = \tau_{yx} , \quad \tau_{xz} = \tau_{zx} , \quad \tau_{yz} = \tau_{zy} . \tag{11.12}$$

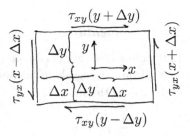

Fig. 11.6   Components of traction.

Consequently, there are only six components of the stress that are independent. It will be convenient to manipulate these six components as a vector (as opposed to a tensor)

$$[\boldsymbol{\sigma}] = [\sigma_x, \sigma_y, \sigma_z, \tau_{xy}, \tau_{xz}, \tau_{yz}]^T . \tag{11.13}$$

Equation (11.8) may be rewritten in terms of the stress vector $\boldsymbol{\sigma}$ as

$$t = \mathcal{P}_{\boldsymbol{n}}\boldsymbol{\sigma} , \tag{11.14}$$

where the matrix "**vector-stress vector dot product**" operator is defined as

$$\mathcal{P}_{\boldsymbol{n}} = \begin{bmatrix} n_x & 0 & 0 & n_y & n_z & 0 \\ 0 & n_y & 0 & n_x & 0 & n_z \\ 0 & 0 & n_z & 0 & n_x & n_y \end{bmatrix} . \tag{11.15}$$

Equation (11.14) may be used in a variety of ways: any of the three quantities may be given, which would then for another quantity being fixed produce the third as the result. Most useful are these two possibilities: t given, produce the stress vector in dependence on the normal; and $\boldsymbol{\sigma}$ given, produce the surface tractions for various normals.

11.3   Local equilibrium

In complete analogy to the model of heat conduction, the global balance equation (11.4) (in this case, balance of linear momentum, for the heat conduction it was balance of heat energy (5.4)) needs to be converted to a local form. The local form expresses dynamic equilibrium of an infinitesimal particle as an equation that holds at a point.

11.3.1   *Change of linear momentum*

There are three terms in the global balance (11.4), and to produce the local form we'll have to convert all three integrals to volume integrals. The first one involves the time derivative of the integral

$$\frac{\mathrm{d}}{\mathrm{d}t} \int_m \boldsymbol{v} \, \mathrm{d}m \ .$$

However, that causes no difficulties since the mass m inside the volume V does not change with time. Therefore,

$$\frac{\mathrm{d}}{\mathrm{d}t} \int_m \boldsymbol{v} \, \mathrm{d}m = \int_m \frac{\mathrm{d}\boldsymbol{v}}{\mathrm{d}t} \, \mathrm{d}m \ . \tag{11.16}$$

Introducing the **mass density** ρ (which as mass per unit volume depends on the deformation, and hence varies with time), we may write $\mathrm{d}m = \rho \mathrm{d}V$ and

$$\int_m \frac{\mathrm{d}\boldsymbol{v}}{\mathrm{d}t} \, \mathrm{d}m = \int_V \frac{\mathrm{d}\boldsymbol{v}}{\mathrm{d}t} \rho \, \mathrm{d}V \ . \tag{11.17}$$

11.3.2   *Stress divergence*

The divergence theorem may be now applied to the third term in (11.4), that is to the surface integral. However, if we introduce the abstract symbol for the divergence of stress by spelling out its definition, we generate more questions than answers. Therefore, it will be instructive to get to the needed form of the divergence theorem in a roundabout way.

   Consider a small volume (parallelepiped) with faces parallel to coordinate planes of the global Cartesian basis (Fig. 11.7, and refer also to Fig. 11.1); for simplicity, the box is drawn as two-dimensional, and it is drawn twice so that we can display the normal and the shear stresses separately. The center of the box is at x, y, z, and the stress components may be expanded into a truncated Taylor series. For instance,

$$\sigma_x(x + \xi\Delta x, y + \eta\Delta y, z + \zeta\Delta z) \approx \sigma_x(x, y, z) + \frac{\partial \sigma_x(x, y, z)}{\partial x}\xi\Delta x$$
$$+ \frac{\partial \sigma_x(x, y, z)}{\partial y}\eta\Delta y + \frac{\partial \sigma_x(x, y, z)}{\partial z}\zeta\Delta z ,$$

where $-1 \le \xi \le +1$, $-1 \le \eta \le +1$, and $-1 \le \zeta \le +1$.

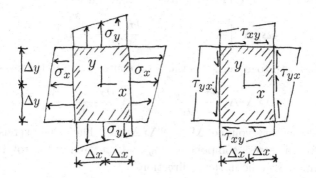

Fig. 11.7   Components of traction.

The box is loaded only by the tractions on its boundary, there are no body loads. Equilibrium in the x-direction requires integration of the stress σ_x over the vertical sides of the box, τ_{xy} over the horizontal sides, and τ_{xz} over the faces parallel to the plane of the paper. For instance, integrating σ_x over the side at $\xi = 1$ leads to

$$\Delta y \Delta z \int_{-1}^{+1} \int_{-1}^{+1} \sigma_x(x + \Delta x, y + \eta\Delta y, z + \zeta\Delta z) \, \mathrm{d}\eta\mathrm{d}\zeta \approx$$
$$\Delta y \Delta z \int_{-1}^{+1} \int_{-1}^{+1} \left[\sigma_x(x, y, z) + \frac{\partial \sigma_x(x, y, z)}{\partial x}\Delta x \right.$$
$$\left. + \frac{\partial \sigma_x(x, y, z)}{\partial y}\eta\Delta y + \frac{\partial \sigma_x(x, y, z)}{\partial z}\zeta\Delta z \right] \, \mathrm{d}\eta\mathrm{d}\zeta .$$

The terms with η and ζ integrate to zero, and the result is

$$4\Delta y \Delta z \left[\sigma_x(x, y, z) + \frac{\partial \sigma_x(x, y, z)}{\partial x}\Delta x \right] .$$

Next, integrating σ_x over the side at $\xi = -1$ leads to

$$\Delta y \Delta z \int_{-1}^{+1} \int_{-1}^{+1} -\sigma_x(x + \Delta x, y + \eta \Delta y, z + \zeta \Delta z) \, d\eta d\zeta \approx$$

$$\Delta y \Delta z \int_{-1}^{+1} \int_{-1}^{+1} \left[-\sigma_x(x, y, z) + \frac{\partial \sigma_x(x, y, z)}{\partial x} \Delta x \right.$$

$$\left. - \frac{\partial \sigma_x(x, y, z)}{\partial y} \eta \Delta y - \frac{\partial \sigma_x(x, y, z)}{\partial z} \zeta \Delta z \right] d\eta d\zeta \, .$$

The terms with η and ζ integrate to zero, and the result is

$$4 \Delta y \Delta z \left[-\sigma_x(x, y, z) + \frac{\partial \sigma_x(x, y, z)}{\partial x} \Delta x \right] \, .$$

Adding these two together gives the total contribution of the stress σ_x as

$$8 \Delta x \Delta y \Delta z \frac{\partial \sigma_x(x, y, z)}{\partial x} = \Delta V \frac{\partial \sigma_x(x, y, z)}{\partial x} \, ,$$

with the elementary volume $\Delta V = 8 \Delta x \Delta y \Delta z$. The same exercise is now repeated for the stress components τ_{xy} and τ_{xz}, giving the total force on the elementary volume in the x-direction

$$\Delta b_x^* = \Delta V \left[\frac{\partial \sigma_x(x, y, z)}{\partial x} + \frac{\partial \tau_{xy}(x, y, z)}{\partial y} + \frac{\partial \tau_{xz}(x, y, z)}{\partial z} \right] \, , \qquad (11.18)$$

and analogously in the other two directions

$$\Delta b_y^* = \Delta V \left[\frac{\partial \tau_{yx}(x, y, z)}{\partial x} + \frac{\partial \sigma_y(x, y, z)}{\partial y} + \frac{\partial \tau_{yz}(x, y, z)}{\partial z} \right] \, , \qquad (11.19)$$

and

$$\Delta b_z^* = \Delta V \left[\frac{\partial \tau_{zx}(x, y, z)}{\partial x} + \frac{\partial \tau_{zy}(x, y, z)}{\partial y} + \frac{\partial \sigma_z(x, y, z)}{\partial z} \right] \, . \qquad (11.20)$$

Now the same argument that was established around Eq. (11.2) will be pursued: put together the total force on the body by collecting the contributions from all the elementary volumes. This can be done in two ways:

(1) Add up all the tractions on the bounding faces of the elementary volumes. The tractions on the shared faces (internal surfaces) will cancel; only the tractions on the exterior surface will be left:

$$\int_S t \, dS$$

(2) Add up all the resultant equivalent volume forces (11.18-11.20), which in the limit will become a volume integral

$$\int_V \boldsymbol{b}^* \, \mathrm{d}V$$

where the imaginary force $\boldsymbol{b}^*$ has components on the Cartesian basis

$$[\boldsymbol{b}^*] = \begin{bmatrix} \dfrac{\partial \sigma_x(x,y,z)}{\partial x} + \dfrac{\partial \tau_{xy}(x,y,z)}{\partial y} + \dfrac{\partial \tau_{xz}(x,y,z)}{\partial z} \\[2ex] \dfrac{\partial \tau_{yx}(x,y,z)}{\partial x} + \dfrac{\partial \sigma_y(x,y,z)}{\partial y} + \dfrac{\partial \tau_{yz}(x,y,z)}{\partial z} \\[2ex] \dfrac{\partial \tau_{zx}(x,y,z)}{\partial x} + \dfrac{\partial \tau_{zy}(x,y,z)}{\partial y} + \dfrac{\partial \sigma_z(x,y,z)}{\partial z} \end{bmatrix} \quad (11.21)$$

and may be recognized as the **stress divergence.**

These two forces are equal, and we have the following form of the divergence theorem

$$\int_V \boldsymbol{b}^* \, \mathrm{d}V = \int_S \boldsymbol{t} \, \mathrm{d}S \; .$$

Using the template of the "vector-stress vector dot product" operator (11.15), we may write the stress divergence as

$$\boldsymbol{b}^* = \mathcal{B}^T \boldsymbol{\sigma} \; , \quad (11.22)$$

where the **stress-divergence operator** $\mathcal{B}^T$ is defined as

$$\mathcal{B}^T = \begin{bmatrix} \partial/\partial x & 0 & 0 & \partial/\partial y & \partial/\partial z & 0 \\ 0 & \partial/\partial y & 0 & \partial/\partial x & 0 & \partial/\partial z \\ 0 & 0 & \partial/\partial z & 0 & \partial/\partial x & \partial/\partial y \end{bmatrix} \; . \quad (11.23)$$

This operator (un-transposed) will make its appearance shortly yet again as the *symmetric gradient operator* to produce strains out of displacements. Using the definitions of both of these useful operators, the **divergence theorem** may be written in terms of stress as

$$\int_V \mathcal{B}^T \boldsymbol{\sigma} \, \mathrm{d}V = \int_S \mathcal{P}\boldsymbol{n}\boldsymbol{\sigma} \, \mathrm{d}S \; . \quad (11.24)$$

11.3.3   *All together now*

Putting the three integrals from (11.4) into the volume-integral form leads to a pointwise expression of local equilibrium (following exactly the same argument as in Section 5.1):

$$\int_V \rho \frac{\mathrm{d}\boldsymbol{v}}{\mathrm{d}t}\,\mathrm{d}V = \int_V \overline{\boldsymbol{b}}\,\mathrm{d}V + \int_V \mathcal{B}^T \boldsymbol{\sigma}\,\mathrm{d}V \quad \Rightarrow \quad \rho \frac{\mathrm{d}\boldsymbol{v}}{\mathrm{d}t} = \overline{\boldsymbol{b}} + \mathcal{B}^T \boldsymbol{\sigma}\ . \quad (11.25)$$

This is a statement of **dynamic equilibrium** of a point particle: On the left-hand side we have the inertial force (mass times acceleration), on the right-hand side is the body load and the force generated by a stress gradient across the particle. Analogously to the heat conduction problem, this local balance equation contains too many variables. The stress plays the role of the heat flux, and it also will be replaced by reference to measurable variables – the strains.

11.4   Strains and displacements

The strains measure the *relative* deformation, and based on the effect they represent when expressed in Cartesian coordinates, they may be divided into two groups: the normal strains (stretches), and the shear strains.

    The strains are an expression of the local variations in the positions of material particles after deformation. The deformation (motion) is expressed as displacements. The **displacement u** is expressed in the Cartesian co-ordinates by components, and connects the locations of a given **material point** (particle) A before deformation and after deformation

$$[\boldsymbol{u}(A,t)] = \begin{bmatrix} x(A,t) \\ y(A,t) \\ z(A,t) \end{bmatrix} - \begin{bmatrix} x(A,0) \\ y(A,0) \\ z(A,0) \end{bmatrix} \ . \quad (11.26)$$

    It will be useful to approach the meaning of strains from the point of view of what happens to tangents to material curves during the deformation. A **material curve** consists of the same material points (particles) at any point in time. A visual may be useful: recall that some specimens have a square grid etched upon them before they are being mechanically tested (deformed). The etching curves that go in one direction may be thought of as sets of points whose one coordinate changes and the other is being held fixed. Figure 11.8 shows a blob of material with two material curves before and after deformation. Before deformation, the curve that is horizontal

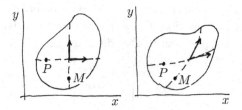

Fig. 11.8 Material curves, and tangents to material curves. Left: before deformation, right: after deformation.

consists of points P such that the coordinates are

$$[P] = \begin{bmatrix} x \\ y = \text{constant} \end{bmatrix},$$

and the curve that is vertical consists of points M such that

$$[M] = \begin{bmatrix} x = \text{constant} \\ y \end{bmatrix}.$$

The parameter that varies along the curve through the point P is x. Therefore, the tangent vector to this curve is

$$\frac{\partial}{\partial x}[P] = \begin{bmatrix} 1 \\ 0 \end{bmatrix}. \tag{11.27}$$

The parameter that varies along the curve through the point M is y. Therefore, the tangent vector to this curve is

$$\frac{\partial}{\partial y}[M] = \begin{bmatrix} 0 \\ 1 \end{bmatrix}. \tag{11.28}$$

The tangent vectors (11.27) and (11.28) are of course just the basis vectors of the Cartesian coordinates.

After deformation, the curve that used to be horizontal consists of points P such that

$$[P] = \begin{bmatrix} x + u_x \\ (y = \text{constant}) + u_y \end{bmatrix},$$

and the curve that is vertical consists of points M such that

$$[M] = \begin{bmatrix} (x = \text{constant}) + u_x \\ y + u_y \end{bmatrix}.$$

Since these are material curves, they are still parameterized by the same parameters as before deformation. Consequently, for the originally horizontal curve we have the tangent vector after deformation

$$\frac{\partial}{\partial x}[P] = \begin{bmatrix} 1 + \dfrac{\partial u_x}{\partial x} \\[2mm] \dfrac{\partial u_y}{\partial x} \end{bmatrix} . \tag{11.29}$$

The parameter that varies along the curve through the point M is y. Therefore, after deformation the tangent vector to this curve is

$$\frac{\partial}{\partial y}[M] = \begin{bmatrix} \dfrac{\partial u_x}{\partial y} \\[2mm] 1 + \dfrac{\partial u_y}{\partial y} \end{bmatrix} . \tag{11.30}$$

The **stretches** measure the relative change in length of the tangent vectors at the same material point before and after deformation. For instance, the tangent vector (11.27) is of unit length before deformation, and the vector (11.29) is of length

$$\sqrt{(1 + \frac{\partial u_x}{\partial x})^2 + (\frac{\partial u_y}{\partial x})^2} = \sqrt{1 + 2\frac{\partial u_x}{\partial x} + (\frac{\partial u_x}{\partial x})^2 + (\frac{\partial u_y}{\partial x})^2} .$$

If we now make the *assumption that the derivatives of the displacement components are very small* in magnitude,

$$|\frac{\partial u_k}{\partial j}| \ll 1, \quad k, j = x, y, z , \tag{11.31}$$

the length of the tangent vector may be expressed as

$$\sqrt{1 + 2\frac{\partial u_x}{\partial x} + (\frac{\partial u_x}{\partial x})^2 + (\frac{\partial u_y}{\partial x})^2} \approx 1 + \frac{\partial u_x}{\partial x} ,$$

and the relative change in length (the stretch in the x direction) is

$$\frac{1 + \dfrac{\partial u_x}{\partial x} - 1}{1} = \epsilon_x .$$

The **shears** measure the change in the angle between originally perpendicular directions of pairs of the Cartesian axes. Therefore, we could measure the change in the angle between the tangents of two intersecting material curves before and after deformation. For the two curves in

Fig. 11.8, the initial angle is $\pi/2$; the cosine of the angle after the deformation is

$$\frac{\partial}{\partial x}[P]^T \frac{\partial}{\partial y}[M] = (1 + \frac{\partial u_x}{\partial x})\frac{\partial u_x}{\partial y} + \frac{\partial u_y}{\partial x}(1 + \frac{\partial u_y}{\partial y})$$

which, again using the assumption (11.31), gives for the change of the angle

$$\frac{\partial u_x}{\partial y} + \frac{\partial u_y}{\partial x} = \gamma_{xy} \ .$$

In this way we define all six strain components: three stretches, and three shears. In fact, we could have defined nine strains (components of the **strain tensor**), which would correspond to the nine components of the Cauchy stress tensor. However, we will stick to the vector representation in this book.

The six strain components are a mixture of the derivatives of the displacement components, and may be expressed in an operator equation, using the definition (11.22)

$$\epsilon = \mathcal{B}u \ , \tag{11.32}$$

where $\mathcal{B}$ is called the **symmetric gradient** (or strain-displacement) operator.

11.5 Constitutive equation

The stress may now be replaced in the balance equation (11.25) by reference to the primary variable, the displacement. However, first we need to discuss the link between the measurable quantities, the strains, and the mathematical device in the balance equation, the stress. As for the thermal model, this link is the constitutive equation.

Since the angular momentum balance (11.12) reduces the number of stress components to six, correspondingly there are six components of strain. The energy of deformation may be defined as the work of each stress component on the corresponding strain component. Let us consider some pre-existing stressed state in a very small neighborhood of a given point. So that we don't have to specify the volume, we will refer to **energy density** (the energy in a certain volume may be obtained by integrating the energy density over this volume). The state of stress is described by the stress vector σ. Let us superimpose an infinitesimal strain variation $d\epsilon$

upon the extant strains. The density of the work of the current stress on the strain change is expressed as

$$d\epsilon^T \sigma \ . \tag{11.33}$$

The constitutive equation that will be of interest in this book is the model of *linear elasticity*. It is expressed as a linear relationship between the strain and the stress, and since these are vectors, the linear relationship, the *constitutive equation*, is expressed as a matrix product

$$\sigma = D\epsilon \ , \tag{11.34}$$

where D is a constant 6×6 matrix of the elastic coefficients (also known as the elasticities); D may be also referred to as the *material stiffness* matrix. Clearly, when there is no strain, the stress is zero. Let us now increase the strain from zero to its final value, ϵ, by scaling with a number $0 \le \theta \le 1$

$$\widehat{\epsilon} = \theta\epsilon \ ,$$

and furthermore use the linear elasticity (11.34). The expression for the change of the energy of deformation density (11.33) will become

$$d\widehat{\epsilon}^T \widehat{\sigma} = d\widehat{\epsilon}^T D\widehat{\epsilon} = d\theta\epsilon^T D\theta\epsilon \ . \tag{11.35}$$

The deformation process starts at $\theta = 0$ and reaches its final stage at $\theta = 1$. In this process, the total energy density stored in the material is

$$\phi(\epsilon) = \int_0^1 d\theta\epsilon^T D\theta\epsilon = \frac{1}{2}\epsilon^T D\epsilon \ . \tag{11.36}$$

Mathematically, the expression $\frac{1}{2}\epsilon^T D\epsilon$ is known as a *quadratic form.* One interesting property of the quadratic form is that the unsymmetrical part of the matrix D does not contribute to the energy:

$$\frac{1}{2}\epsilon^T D\epsilon = \left(\frac{1}{2}\epsilon^T D\epsilon\right)^T = \frac{1}{2}\epsilon^T D^T \epsilon \quad \Rightarrow \quad \frac{1}{2}\epsilon^T (D - D^T)\epsilon = 0 \ .$$

Because the energy of deformation is a fundamental quantity, from physical principles, and from the point of view of mathematical modeling, this is a very good reason for postulating a priori the *symmetry of the material stiffness, $D = D^T$.*

At the moment, we will leave the material stiffness matrix unspecified, since a detailed discussion follows in Section 12.5.

11.6 Boundary conditions

Similarly to the heat conduction problem, at each point on the bounding surface a boundary condition is required. The boundary conditions may be in terms of the primary variable, the displacement, or in terms of the flux variable, the stress. For heat conduction, the boundary condition in terms of flux referred to the *normal* flux only, since the flux parallel to the surface is essentially impossible to control in physical experiments. Similarly, for elasticity the flux boundary condition will not attempt to prescribe all six components of stress, but rather the "projection" of stress, the traction.

A complicating circumstance is that the primary variable and the traction both have three components. Therefore, the surface of the solid needs to be considered three times as to the appropriate boundary condition, *once for each component*.

Selection of the appropriate boundary conditions is critical to successful modeling. Typically, the boundary conditions that are applied to our models are only approximations of the physical reality. Thus, the first guidelines for the application of boundary conditions will be based on *physical considerations*.

11.6.1 *Example: concrete dam*

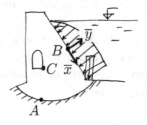

Fig. 11.9   Example of boundary conditions: concrete dam with a tunnel.

In the first example we will consider a concrete dam. Figure 11.9 shows the cross-section of a dam, and of interest is the stress near point C in the corner of the tunnel. Therefore, we could decide to neglect the deformation of the soil near the base of the dam, and prescribe zero magnitude for all displacement components along the surface A. In reality, this is not strictly true, and a so-called *modeling error* is being introduced by making this choice. In a careful analysis, the influence of this error would be assessed, for instance by varying the boundary condition, or including the soil in the analysis.

On the surfaces exposed to the water behind the dam, including the one with point A, the structure is loaded by water *pressure*, which is a special kind of traction: using an ad hoc Cartesian coordinate system as indicated in the figure, the traction components are

$$t_{\bar{x}} = 0, \quad t_{\bar{y}} = p, \quad t_{\bar{z}} = 0$$

where p is the water pressure at the particular location.

All the other surfaces in the model that show up as curves, are assigned the so-called **traction-free** boundary condition: there are no known loads applied there. Furthermore, the model may be formulated using just two coordinates (reduced model of the so-called "plane strain" type): the remaining surfaces that are parallel to the plane of the paper will be assigned zero displacement normal to the paper, and zero shear components of traction in the plane of the paper. This type of model is discussed in detail later in the textbook.

Let us now examine the associated variable (so-called work-conjugate variable)– traction – along the surfaces where we prescribed displacements, for instance at point A. The physical meaning of such tractions, which are generated in the soil by the stress in the bulk of the dam near the surface, is clear: they are the **reactions**. They are initially unknown, but as soon as the displacements are available from the solution, the reactions may be calculated.

The work-conjugate variable along the traction-free surfaces is displacement, which is initially unknown, but which will be produced during the solution process. Similarly, displacement is unknown on the surfaces exposed to the water behind the dam.

11.6.2 *Example: rigid punch*

To model deformation under a stiff punch which is vertically pushed against a block of material, we may apply a set of approximate bilateral boundary conditions. (A further refinement would be a unilateral, contact, condition. But this is out of the range of this book.) Firstly, the punch may be assumed completely rigid, and perfect contact of the punch with the material underneath may be assumed. Also, perfect sticking or perfect slip under the punch may be assumed: the former when the surfaces in contact have a very high coefficient of friction, or perhaps they're bonded, the latter when

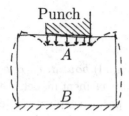

Fig. 11.10 Example of boundary conditions: rigid punch.

the surfaces are lubricated. For perfect stick, we could prescribe the motion of the points on the conduct surface to be entirely driven by the punch: only vertical displacement, zero horizontal displacement. For perfect slip, the vertical motion is prescribed to be that of the punch, but a horizontal displacement under the punch is free. Therefore, for perfect slip we would apply the condition of zero shear traction under the punch.

11.6.3   *Formal definition of the boundary conditions*

At each point of the boundary we define a Cartesian coordinate system. It could be the surface-aligned system of Fig. 11.3, it could be the global system, or an arbitrarily oriented system. For instance, refer to Fig. 11.11 where each of the surfaces has its own coordinate system.

Both the traction vector and the displacement vector at a given point may be written in such a coordinate system in terms of the components as

$$[\boldsymbol{t}] = \begin{bmatrix} t_x \\ t_y \\ t_z \end{bmatrix} \quad , \quad [\boldsymbol{u}] = \begin{bmatrix} u_x \\ u_y \\ u_z \end{bmatrix} \ .$$

For each direction $i = x, y, z$ we separate the surface S into two disjoint parts:

(1) $S_{t,i}$ where the traction component t_i is being prescribed;
(2) $S_{u,i}$ where the displacement component u_i is being prescribed.

It holds that $S = S_{t,i} \bigcup S_{u,i}$, and it could be that either $S_{t,i} = \emptyset$ or $S_{u,i} = \emptyset$.
The boundary conditions may now be expressed as

$$t_i = (\mathcal{P}\boldsymbol{n}\boldsymbol{\sigma})_i = \bar{t}_i \quad \text{on } S_{t,i} \quad \text{for } i = x, y, z \qquad (11.37)$$

as the **traction** (natural) **boundary condition** for the ith component,

and

$$u_i = \overline{u}_i \quad \text{on } S_{u,i} \quad \text{for } i = x, y, z \qquad (11.38)$$

as the **displacement** (essential) **boundary condition** for the ith component. When setting up a finite element model, note that the natural boundary condition need only be specified explicitly when $\overline{t}_i \neq 0$; the case of $\overline{t}_i = 0$ is implicitly active when we say nothing. In particular, no boundary condition needs to be explicitly defined for any component for a traction-free surface.

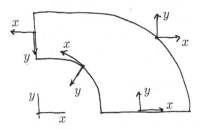

Fig. 11.11   Local coordinate systems used for boundary condition definitions.

11.6.4   *Inadmissible "concentrated" boundary conditions*

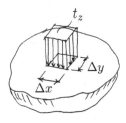

Fig. 11.12   Concentrated force as the limit of traction on infinitesimally small area.

Consider that a resultant force of magnitude F is to be applied along the z-direction, that is perpendicularly to the surface shown in Fig. 11.12, as traction t_z applied to the area $\Delta x \Delta y$. As we need to have

$$F = t_z \Delta x \Delta y ,$$

if $\Delta x \to 0, \Delta y \to 0$, the traction component must approach infinity $t_z \to \infty$. In addition, since from the boundary conditions we have $\sigma_z = t_z$, we must conclude that in the immediate vicinity of the infinitesimal patch on the surface, at least some of the stresses must approach infinity as the traction component approaches infinity. The problem of a force applied to an infinite half space has been solved analytically by Boussinesq and others [Sokolnikoff (1983)], and perhaps the most significant conclusion is that the displacement under the force is infinite. Consequently, for any finite force, the *energy* in the system is *infinite*. As a consequence, we should remember the following caveats when using a *concentrated force as a boundary condition*:

(1) Trying to obtain a converged solution for the displacement under the force or for the energy is pointless;
(2) Displacements and stresses near the point of application of the force are most likely wrong for any purpose;
(3) Displacements or stresses removed from the point of application of the force may be useful, but we have to always ask ourselves whether the concentrated force is truly needed or whether we use it only because we haven't thought the problem through.

Furthermore, as a corollary, we must conclude that if we apply a *displacement boundary condition at a point*, the associated reaction will be zero in the limit, which is wrong, unless we can guarantee for instance from global force equilibrium conditions that the reaction should be zero.

Very similar analysis may be performed for a *distributed load along a curve*: Figure 11.13. To maintain a finite value of the distributed load (force per unit length) as $\Delta y \to 0$, the traction t_z must approach infinity. The same list of caveats applies.

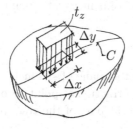

Fig. 11.13   Distributed load along a curve as the limit of traction on infinitesimally small area.

To summarize, the following should be remembered for concentrated force boundary conditions:

Do not use the concentrated force or force along a curve boundary condition *unless* it is essential. Remember the caveats.

Furthermore, this should be remembered for concentrated displacement boundary conditions (support at a point, or support along a curve):

Do not use the concentrated support boundary condition *unless* the associated reaction is guaranteed to be zero.

11.6.5 *Symmetry and anti-symmetry*

Considerable benefits may be often derived when the solution is expected to possess either symmetry, or anti-symmetry.

For the solution to display **symmetry with respect to reflection** in a symmetry plane, all the ingredients that go into the definition of the problem must display the same kind of symmetry: the geometry, the material, the boundary conditions, the initial conditions.

Let us first look at the conditions that must hold for the *displacements* on the plane of symmetry: Figure 11.14. By inspection, we see that the arrow representing displacement at point P is reflected into an arrow at point P' which is best described in components which are (i) in the plane of symmetry: these are the same for both arrows; and (ii) perpendicular to the plane of symmetry: these have opposite signs. Therefore, if P is made to approach the plane of symmetry, its mirror image merges with it when they both reach the plane of symmetry (point M), and since the two displacements then must be the same, we may conclude that the perpendicular component of displacement $u_\perp$ at a point on the plane of symmetry must be zero

$$u_\perp = 0 \,. \tag{11.39}$$

Now for the *tractions* on the plane of symmetry. By the symmetry conditions, the shear part of the traction at point M on the surface with normal $\boldsymbol{n}$ must be equal to the shear part of the traction on the surface with the opposite normal $-\boldsymbol{n}$, i.e.

$$\boldsymbol{t}_{s(-n)} = \boldsymbol{t}_{s(n)} \,.$$

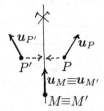

Fig. 11.14 Symmetric displacement pattern.

Furthermore, using equation (11.7) we have

$$t_{s(n)} = t_{(n)} - \left(n \cdot t_{(n)}\right) n \, ,$$

and substituting from (11.15) with σ being the stress at the point M (which is the same irrespectively of the normal)

$$t_{s(n)} = \mathcal{P}_n \sigma - (n \cdot \mathcal{P}_n \sigma) n \, ,$$

$$t_{s(-n)} = \mathcal{P}_{-n} \sigma - (-n \cdot \mathcal{P}_{-n} \sigma)(-n) = -\mathcal{P}_n \sigma + (n \cdot \mathcal{P}_n \sigma) n = -t_{s(n)} \, .$$

In order for both requirements to be satisfied,

$$t_{s(n)} = 0 \, , \qquad (11.40)$$

must hold.

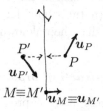

Fig. 11.15 Anti-symmetric displacement pattern.

The state of ***anti-symmetry with respect to reflection in a plane*** is defined by conditions that specify it as the opposite of symmetry. The situation is illustrated in Fig. 11.15. The arrow representing *displacement* at point P is transformed into an arrow at point P' which in terms of components gives (i) opposite sign parallel with the plane of anti-symmetry; and (ii) same sign in the direction perpendicular to the plane of anti-symmetry.

Table 11.1 Boundary conditions on the plane of symmetry or anti-symmetry

Quantity	Symmetry	Anti-symmetry
Tractions $\parallel$	0	unknown
Displacements $\parallel$	unknown	$\boxed{0}$
Traction $\perp$	unknown	0
Displacement $\perp$	$\boxed{0}$	unknown

Therefore, when P is made to approach the plane of symmetry and its mirror image merges with it at point M, we conclude that the two components of displacement $u_{\parallel,i}$ at a point on the plane of symmetry must be zero

$$u_{\parallel,i} = 0 \quad \text{for the two in-plane directions } i \; . \qquad (11.41)$$

For the tractions, an analysis quite similar to that leading to Eq. (11.40), but applied to the normal component of the traction, leads to the condition

$$t_\perp = 0 \; . \qquad (11.42)$$

Therefore, we can summarize the boundary conditions on the plane of symmetry or on the plane of anti-symmetry with the delightfully simple Table 11.1. The boundary conditions that need to be explicitly prescribed in finite element analyses are boxed in the Table 11.1: zero tractions are incorporated automatically (natural boundary conditions!) and need not be explicitly specified. Note that the unknown tractions are the reactions.

11.6.6 *Example: a pure-traction problem*

Sometimes we encounter stress analysis problems where only a statically equilibrated set of traction and/or body loads is given. As an example, we consider the dog bone tensile specimen of Fig. 11.16 (slice through the axis of symmetry is shown). Uniform tractions are applied at the opposite cross-sections, equal in magnitude, but of opposite sign, so that the specimen is in static equilibrium: the so-called ***pure-traction problem***. As such, this set of boundary conditions does not allow for the finite element solution to be computed without additional devices: the entire specimen may be translated or rotated as a rigid body without any change in the stress state. Therefore, the stiffness matrix of the structure as a whole is singular.

Fig. 11.16   Example of boundary conditions: dog bone specimen under tension.

The rigid body motion may be described by displacements of the form

$$
\begin{bmatrix} u_x \\ u_y \\ u_z \end{bmatrix} = \begin{bmatrix} 0 & -\Theta_z & \Theta_y \\ \Theta_z & 0 & -\Theta_x \\ -\Theta_y & \Theta_x & 0 \end{bmatrix} \begin{bmatrix} x \\ y \\ z \end{bmatrix} + \begin{bmatrix} a_x \\ a_y \\ a_z \end{bmatrix} , \qquad (11.43)
$$

where $\Theta_x, \Theta_y, \Theta_z$ and a_x, a_y, a_z are constants describing **rotation** (through the skew-symmetric matrix – another way of writing a cross product of two vectors), and **translation** of the points of a rigid body.

It doesn't take too much effort to verify that strains computed from displacements (11.43) are all identically zero. Therefore, also the deformation energy induced by the rigid body motion is zero, as is easily verified by referring to (11.36). As shown in Section 12.9, the global stiffness matrix of a structure that can move as a rigid body is by necessity singular. To restore the full rank of the stiffness matrix, all possible rigid body modes must be prevented by additional supports (displacement boundary conditions). Taking advantage of any symmetry conditions is a big help. Even though we may not be willing to actually solve the problem on a quarter of the geometry (producing a separate CAD model for each analysis is often not convenient), just inserting features such as split lines to which symmetry conditions may be applied will do the trick. For instance, Fig. 11.17 illustrates these two possibilities: on the left, one-quarter model is extracted from the full geometry, with zero normal displacement on the cut planes; on the right, split lines have been inserted, to which zero displacement perpendicular to the corresponding symmetry plane will be applied. Note that this complies with the modeling rule of Section 11.6.4: since the tractions applied at the end cross-sections are self-balancing, we can prove that all reactions along the symmetry planes must be zero. Therefore, it is okay to apply an "inadmissible" support along a curve.

Another possibility is to support the specimen by applying six point supports. These would be selected to (i) prevent any rigid body motion, while (ii) ensuring all reactions at these point supports were identically zero. This would be achieved by formulating six equilibrium conditions for the specimen as a rigid body, and checking that the reactions at the supports

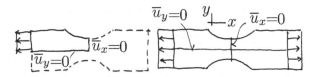

Fig. 11.17   Dog bone specimen under tension with two ways of using symmetry.

would make equilibrium possible and unique (even though they *all must be zero*). To ensure that the reactions are statically determinate, the distances between the point supports must be able to change freely: while we want to support the specimen as a rigid body, it is not rigid, it needs to freely deform. An example of possible system of supports is shown in Fig. 11.18.

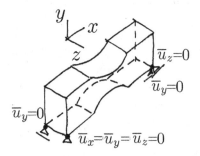

Fig. 11.18   Dog bone specimen supported with point constraints as a rigid body.

11.6.7   *Example: shaft under torsion*

Shear traction components are often generated by frictional contact between interacting bodies. However, contact problems are well outside the scope of this textbook, they are nonlinear and involve inequality constraints. The other situation in which we might wish to apply shear tractions is when we formulate simplified models in which the effect of the omitted part of a structure is introduced as a resultant (force or torque) into the model.

As an example, we will consider a shaft of circular cross-section, with two through-holes. The focus of our interest is the local stress concentration around the holes when the shaft is subjected to a known torque. As we do not wish to model the actual transmission of the torque into the shaft, the geometry of the shaft is reduced to just the small neighborhood of the holes, and the torque generated at the end points is applied as prescribed shear tractions in the end cross-sections of the short stump.

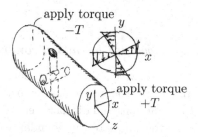

Fig. 11.19   Example of boundary conditions: shaft loaded by torque.

The way in which we distribute the shear tractions is only an approximation of the stress distribution that would exist in the complete part. Based on experimental observations accompanied by analyses of a few particular cases, the so-called *Saint-Venant's principle* [Timoshenko (1983); Barber (1999)], may be invoked: If a set of self-equilibrated tractions is applied on a limited subset of the boundary, its effect will be negligible beyond a certain range. By necessity, this principle is somewhat vague, and its applicability needs to be assessed case-by-case. Figure 11.20 illustrates the meaning: consider a beam of solid section, to which a traction $\bar{t}$ of a nonzero resultant is applied. If the traction is perturbed by a self-equilibrated load $\hat{t}$, the stresses will change significantly only in a region extending in all directions approximately by the characteristic dimension of the beam d.

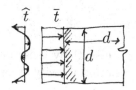

Fig. 11.20   Illustration of Saint-Venant's principle.

Coming back to the shaft: relying on the Saint-Venant's principle, we apply the resultant torque by any convenient distribution of shear tractions. Again, this is a pure-traction problem, so essential boundary conditions to prevent rigid body motion should be added to make the solution unique. In this case, for instance a plane of anti-symmetry exists, or point supports could be added at convenient locations (for instance on the axis of the shaft).

11.6.8   *Example: overspecified boundary conditions*

At each point of the boundary, for each component either displacement or traction must be known. In some special circumstances, it might be of interest to prescribe *both* tractions and displacements, for one or more components, and to compute the values of the boundary conditions elsewhere. Consider for instance the thin plate of Fig. 11.21. The deflection could be measured at the top surface, and the observation that it is traction free could be made. The question then is, could an elasticity problem be solved to determine the unknown tractions on the surfaces with points A, B, C? The answer is yes, however the existence of a solution (any solution!) is not guaranteed at all. In particular, the assumptions on the surface containing point D must be such that the displacements and the tractions along the surface are consistent. In this book we shall not attempt to solve problems of this nature, as special formulations are required.

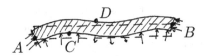

Fig. 11.21   Example of boundary conditions: plate with unknown boundary conditions.

11.7   Initial conditions

The *initial conditions* are essentially those discussed for the taut wire model: prescribe displacement and velocity at each point of the domain.

$$u(x,0) = \bar{W}(x), \quad \frac{\partial u}{\partial t}(x,0) = \bar{V}(x) , \qquad (11.44)$$

where $\bar{W}(x)$ (the initial deflection) and $\bar{V}(x)$ (the initial velocity) are known functions. These functions need to be compatible with the boundary conditions as time $t = 0$: at all points x where we prescribe displacements on the boundary, the functions $\bar{W}(x)$ and $\bar{V}(x)$ must have components of the same value that is being prescribed.

Finally, it needs to be realized that the initial displacement determines also a distribution of the *initial stress*. If the initial displacement is zero everywhere, the strains and hence also the stresses are zero everywhere. On the contrary, a nonzero initial displacement implies initial strains (likely to be also nonzero). From these strains, the initial stresses follow by the constitutive equation. Nonzero initial stresses need to be then applied as an initial condition.

Chapter 12

Galerkin Formulation for Elastodynamics

Rearranging the balance equation (11.25) into the residual form leads to

$$r_B = \rho \frac{d\boldsymbol{v}}{dt} - \bar{\boldsymbol{b}} - \mathcal{B}^T \boldsymbol{\sigma} \,, \tag{12.1}$$

which is a statement of imbalance when the force residual is nonzero. This is in complete analogy to the model of the taut wire, or the model of heat conduction. To arrive at a usable form of the method of weighted residuals for elastodynamics, the plan of action is the same as for the other models.

12.1 Manipulation of the residuals

(1) Formulate the weighted residual equations for the balance equation, the force boundary condition, and the displacement boundary condition.
(2) Satisfy the displacement condition by design of the trial functions: that will subject the trial functions to a condition along parts of the boundary.
(3) Shift the derivatives from the stress to the test function. This will incorporate the natural boundary conditions in the balance residual equation (and eliminate the force boundary condition residual from further consideration); it will also place a condition on the form of the test functions.

12.1.1 The first two steps

We begin with step 1: The natural boundary condition (11.37) leads to the residual

$$r_{t,i} = (\mathcal{P}_n \boldsymbol{\sigma})_i - \bar{t}_i \quad \text{on } S_{t,i} \quad \text{for } i = x, y, z \,, \tag{12.2}$$

which will be incorporated into the balance residual equation, and the displacement boundary condition (11.38) gives the residual

$$r_{u,i} = u_i - \overline{u}_i \quad \text{on } S_{u,i} \quad \text{for } i = x, y, z, \tag{12.3}$$

that will be satisfied by the choice of the trial functions (which takes care of the step 2).

The weighted residual equations are simply integrals of the residuals over the corresponding surface. Since the displacement boundary condition residual is identically zero, it may be ignored. The traction boundary condition residual equation reads

$$\int_{S_{t,i}} r_{t,i} \eta_i \, dS = 0 ,$$

or, expanded,

$$\int_{S_{t,i}} r_{t,i} \eta_i \, dS = \int_{S_{t,i}} \left[(\mathcal{P}_n \boldsymbol{\sigma})_i - \overline{t}_i \right] \eta_i \, dS = 0 . \tag{12.4}$$

The balance weighted residual equation reads

$$\int_V \boldsymbol{r}_B \cdot \boldsymbol{\eta} \, dV = \int_V \boldsymbol{\eta} \cdot \boldsymbol{r}_B \, dV = 0 , \tag{12.5}$$

where $\boldsymbol{\eta}$ is a vector test function (with three components). At this point, we only require that the test function be sufficiently smooth for the integral to exist. The dot product of the residual and the vector test function is written in the "dot" form; when the weighted residual equation is written in terms of the components, transposes must be used as

$$\int_V [\boldsymbol{r}_B]^T [\boldsymbol{\eta}] \, dV = \int_V [\boldsymbol{\eta}]^T [\boldsymbol{r}_B] \, dV = 0 .$$

12.1.2 *Step 3: Preliminaries*

While the first two terms on the right-hand side of (12.1) present no difficulties, the stress term needs to be treated similarly to the previous two models to move one derivative from the stress to the test function. Therefore, in the next few paragraphs we focus on the integral

$$\int_V \boldsymbol{\eta} \cdot \mathcal{B}^T \boldsymbol{\sigma} \, dV .$$

It will be sought as one constituent of the chain-rule result (analogously to Eq. (6.4) for the heat conduction). The inner product of $\boldsymbol{\eta}$ and $\boldsymbol{\sigma}$ may be expressed using the vector-stress vector dot product operator (11.15) in the form $\mathcal{P}\boldsymbol{\eta}\boldsymbol{\sigma}$ (which is a vector). Therefore, we need an identity for the chain rule applied to the divergence div $(\mathcal{P}\boldsymbol{\eta}\boldsymbol{\sigma})$:

$$\operatorname{div}(\mathcal{P}\boldsymbol{\eta}\boldsymbol{\sigma}) = (\mathcal{B}\boldsymbol{\eta}) \cdot \boldsymbol{\sigma} + \boldsymbol{\eta} \cdot \mathcal{B}^T\boldsymbol{\sigma} \,. \tag{12.6}$$

It may not be immediately clear *why* the right-hand side has this form, but to verify this formula is straightforward, albeit tedious. Expressing the rightmost term of (12.6) by the other two, we obtain

$$\int_V \boldsymbol{\eta} \cdot \mathcal{B}^T\boldsymbol{\sigma} \, \mathrm{d}V = \int_V \operatorname{div}(\mathcal{P}\boldsymbol{\eta}\boldsymbol{\sigma}) \, \mathrm{d}V - \int_V (\mathcal{B}\boldsymbol{\eta}) \cdot \boldsymbol{\sigma} \, \mathrm{d}V \,.$$

The divergence theorem (5.10) may be applied to the first term on the right to yield

$$\int_V \boldsymbol{\eta} \cdot \mathcal{B}^T\boldsymbol{\sigma} \, \mathrm{d}V = \int_S (\mathcal{P}\boldsymbol{\eta}\boldsymbol{\sigma}) \cdot \boldsymbol{n} \, \mathrm{d}S - \int_V (\mathcal{B}\boldsymbol{\eta}) \cdot \boldsymbol{\sigma} \, \mathrm{d}V \,.$$

The traction boundary condition (11.37) references $\mathcal{P}\boldsymbol{n}\boldsymbol{\sigma}$. To extricate this form from $(\mathcal{P}\boldsymbol{\eta}\boldsymbol{\sigma}) \cdot \boldsymbol{n}$ we note that the result of this dot product is a scalar (a number). This indicates that the stress is involved in a double dot product: the stress vector is dotted with one vector, which is subsequently dotted with the second vector. Indeed, it is easily verified by multiplying through that

$$(\mathcal{P}\boldsymbol{\eta}\boldsymbol{\sigma}) \cdot \boldsymbol{n} = (\mathcal{P}\boldsymbol{n}\boldsymbol{\sigma}) \cdot \boldsymbol{\eta} \,.$$

As a result we obtain

$$\int_V \boldsymbol{\eta} \cdot \mathcal{B}^T\boldsymbol{\sigma} \, \mathrm{d}V = \int_S \boldsymbol{\eta} \cdot (\mathcal{P}\boldsymbol{n}\boldsymbol{\sigma}) \, \mathrm{d}S - \int_V (\mathcal{B}\boldsymbol{\eta}) \cdot \boldsymbol{\sigma} \, \mathrm{d}V \,.$$

It will be useful to summarize the balance residual equation (12.5) now as

$$\int_V \boldsymbol{\eta} \cdot \rho \frac{\mathrm{d}\boldsymbol{v}}{\mathrm{d}t} \, \mathrm{d}V - \int_V \boldsymbol{\eta} \cdot \overline{\boldsymbol{b}} \, \mathrm{d}V$$
$$- \int_S \boldsymbol{\eta} \cdot (\mathcal{P}\boldsymbol{n}\boldsymbol{\sigma}) \, \mathrm{d}S + \int_V (\mathcal{B}\boldsymbol{\eta}) \cdot \boldsymbol{\sigma} \, \mathrm{d}V = 0 \,. \tag{12.7}$$

12.1.3 *Step 3: The glorious conclusion*

Closely paralleling Section (6.6), the surface will now be split, for each component, into the part where traction is known, and the part where displacement is being prescribed

$$\int_S \boldsymbol{\eta} \cdot (\mathcal{P}\boldsymbol{n}\boldsymbol{\sigma}) \; \mathrm{d}S = \sum_{i=x,y,z} \int_{S_{t,i}} (\boldsymbol{\eta})_i \, (\mathcal{P}\boldsymbol{n}\boldsymbol{\sigma})_i \; \mathrm{d}S + \int_{S_{u,i}} (\boldsymbol{\eta})_i \, (\mathcal{P}\boldsymbol{n}\boldsymbol{\sigma})_i \; \mathrm{d}S \;,$$

where $(\boldsymbol{\eta})_i$ is the ith component of the test function. On the $S_{t,i}$ subset the traction component i is known from (11.37), but on the $S_{u,i}$ subset of the bounding surface, the component i of the traction is not known, it represents the reaction. To be able to ignore the reactions, we will resort to the same trick as for the other PDE models, namely we will put in place the requirement

$$(\boldsymbol{\eta})_i = 0 \quad \text{on} \quad S_{u,i} \;.$$

Therefore, with the constraint on the trial function to satisfy the essential boundary conditions on $S_{u,i}$, and a constraint on the test function to vanish on $S_{u,i}$, we have the weighted balance residual equation

$$\int_V \boldsymbol{\eta} \cdot \rho \frac{\mathrm{d}\boldsymbol{v}}{\mathrm{d}t} \; \mathrm{d}V - \int_V \boldsymbol{\eta} \cdot \overline{\boldsymbol{b}} \; \mathrm{d}V - \sum_{i=x,y,z} \int_{S_{t,i}} (\boldsymbol{\eta})_i \, (\mathcal{P}\boldsymbol{n}\boldsymbol{\sigma})_i \; \mathrm{d}S + \int_V (\mathcal{B}\boldsymbol{\eta}) \cdot \boldsymbol{\sigma} \; \mathrm{d}V = 0$$

$$(12.8)$$

and using the natural boundary condition (12.4), we obtain the **final form of the weighted residual equation**

$$\int_V \boldsymbol{\eta} \cdot \rho \frac{\mathrm{d}\boldsymbol{v}}{\mathrm{d}t} \; \mathrm{d}V - \int_V \boldsymbol{\eta} \cdot \overline{\boldsymbol{b}} \; \mathrm{d}V - \sum_{i=x,y,z} \int_{S_{t,i}} (\boldsymbol{\eta})_i \, \overline{t}_i \; \mathrm{d}S + \int_V (\mathcal{B}\boldsymbol{\eta}) \cdot \boldsymbol{\sigma} \; \mathrm{d}V = 0$$

$$u_i = \overline{u}_i \quad \text{and} \quad (\boldsymbol{\eta})_i = 0 \quad \text{on } S_{u,i} \quad \text{for } i = x, y, z \;.$$

$$(12.9)$$

So far we have been using the velocity and the stress vector for convenience and brevity, but to produce a displacement-based computational model these will have to be replaced by references to the displacement field. Using

$$\boldsymbol{v} = \frac{\mathrm{d}\boldsymbol{u}}{\mathrm{d}t} \quad \Rightarrow \quad \frac{\mathrm{d}\boldsymbol{v}}{\mathrm{d}t} = \frac{\mathrm{d}^2\boldsymbol{u}}{\mathrm{d}t^2} = \ddot{\boldsymbol{u}} \;,$$

and the constitutive equation (11.34) and the displacement-strain relation (11.32), the form that goes into the discretization process reads

$$\int_V \boldsymbol{\eta} \cdot \rho \ddot{\boldsymbol{u}} \; \mathrm{d}V - \int_V \boldsymbol{\eta} \cdot \overline{\boldsymbol{b}} \; \mathrm{d}V - \sum_{i=x,y,z} \int_{S_{t,i}} (\boldsymbol{\eta})_i \, \overline{t}_i \; \mathrm{d}S + \int_V (\mathcal{B}\boldsymbol{\eta}) \cdot \boldsymbol{D}\mathcal{B}\boldsymbol{u} \; \mathrm{d}V = 0$$

$$u_i = \overline{u}_i \quad \text{and} \quad (\boldsymbol{\eta})_i = 0 \quad \text{on } S_{u,i} \quad \text{for } i = x, y, z \;.$$

12.2   Method of weighted residuals as the principle of virtual work

An alternative route to Eq. (12.9) is via the **principle of virtual work**. The test function is interpreted as a **virtual displacement**. The constraint on the test function is postulated a priori: virtual displacement is **kinematically admissible**. The various terms in (12.9) are interpreted as the virtual work of the inertial forces, applied body load, applied tractions, and internal forces.

This approach tends to seem somewhat arbitrary, since a number of concepts are postulated to be taken on faith. In this book we therefore avoid this viewpoint. Nevertheless, it may be useful to be aware of these possible interpretations of the various terms as virtual quantities (especially work).

12.3   Discretizing

The primary variable, the displacement, is a vector quantity

$$\boldsymbol{u} = u_x \boldsymbol{e}_x + u_y \boldsymbol{e}_y + u_z \boldsymbol{e}_z \,,$$

where $u_x, \dots$ are the components, and $\boldsymbol{e}_x, \dots$ are the basis vectors. All computer manipulations are performed in terms of components; the basis vectors are necessary only when a transition needs to be made from one basis to another. For simplicity, in this work and in the toolbox SOFEA, the basis in which the *displacement field* is expressed is the *global Cartesian basis*.

12.3.1   *The trial function*

The trial displacement vector function will be expressed in terms of the components in the global Cartesian basis (the basis is implied) as (compare with Sections 2.7 and 6.5)

$$[\boldsymbol{u}(\boldsymbol{x},t)] = \begin{bmatrix} u_x(\boldsymbol{x},t) \\ u_y(\boldsymbol{x},t) \\ u_z(\boldsymbol{x},t) \end{bmatrix} = \sum_{i=1}^{N} N_i(\boldsymbol{x})[\boldsymbol{u}_i(t)] = \sum_{i=1}^{N} N_i(\boldsymbol{x}) \begin{bmatrix} u_{ix}(t) \\ u_{iy}(t) \\ u_{iz}(t) \end{bmatrix} .$$

$$(12.10)$$

Here $N_i(\boldsymbol{x})$ is a finite element basis function (at this point we assume it is defined on a three-dimensional mesh), and $u_{ix}(t), \dots$ are nodal degrees of freedom (displacements at nodes) as functions of time. As in Section 6.5, it will be useful to separate the free degrees of freedom from the prescribed displacements. However, this will be somewhat more complicated because now we have three components, each of which has different sets of the free degrees of freedom and the prescribed ones. As an illustration consider Fig. 12.1: for the x component, the free degrees of freedom are u_{2x}, the prescribed (at zero value) are u_{1x}, u_{3x}; for the y component, the free degrees of freedom are u_{3y}, the prescribed (at zero value) are u_{1y}, u_{2y}. One possi-

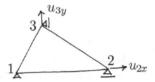

Fig. 12.1   A single element, with free degrees of freedom.

bility is to split the sum and write (note that the sets free i and prescribed i are in general different for each i)

$$\begin{bmatrix} u_x(\boldsymbol{x},t) \\ u_y(\boldsymbol{x},t) \\ u_z(\boldsymbol{x},t) \end{bmatrix} = \begin{bmatrix} \displaystyle\sum_{\text{free } i} N_i(\boldsymbol{x})u_{ix}(t) + \sum_{\text{prescribed } i} N_i(\boldsymbol{x})\overline{u}_{ix}(t) \\ \displaystyle\sum_{\text{free } i} N_i(\boldsymbol{x})u_{iy}(t) + \sum_{\text{prescribed } i} N_i(\boldsymbol{x})\overline{u}_{iy}(t) \\ \displaystyle\sum_{\text{free } i} N_i(\boldsymbol{x})u_{iz}(t) + \sum_{\text{prescribed } i} N_i(\boldsymbol{x})\overline{u}_{iz}(t) \end{bmatrix}, \quad (12.11)$$

but a more convenient approach is the following trick: pretend the node has *both* free and prescribed degrees of freedom for each component at the same time, and zero out those that are inactive

$$\begin{bmatrix} u_x(\boldsymbol{x},t) \\ u_y(\boldsymbol{x},t) \\ u_z(\boldsymbol{x},t) \end{bmatrix} = \sum_{\text{all } i} N_i(\boldsymbol{x}) \left\{ \begin{bmatrix} u_{ix}(t) \\ u_{iy}(t) \\ u_{iz}(t) \end{bmatrix} + \begin{bmatrix} \overline{u}_{ix}(t) \\ \overline{u}_{iy}(t) \\ \overline{u}_{iz}(t) \end{bmatrix} \right\}, \quad (12.12)$$

where we define

$u_{ix}(t) = 0$ if the x degree of freedom at node i is prescribed;
$u_{iy}(t) = 0$ if the y degree of freedom at node i is prescribed;
$u_{iz}(t) = 0$ if the z degree of freedom at node i is prescribed,

and

$$\overline{u}_{ix}(t) = 0 \text{ if the } x \text{ degree of freedom at node } i \text{ is free;}$$
$$\overline{u}_{iy}(t) = 0 \text{ if the } y \text{ degree of freedom at node } i \text{ is free;}$$
$$\overline{u}_{iz}(t) = 0 \text{ if the } z \text{ degree of freedom at node } i \text{ is free.}$$

Thus, for the example of Fig. 12.1 we have the following

$$\begin{bmatrix} u_{1x}(t) \\ u_{1y}(t) \end{bmatrix} = \begin{bmatrix} 0 \\ 0 \end{bmatrix}, \quad \begin{bmatrix} \overline{u}_{1x}(t) \\ \overline{u}_{1y}(t) \end{bmatrix} = \begin{bmatrix} \text{as given} \\ \text{as given} \end{bmatrix},$$

for node 1,

$$\begin{bmatrix} u_{2x}(t) \\ u_{2y}(t) \end{bmatrix} = \begin{bmatrix} u_{2x}(t) \\ 0 \end{bmatrix}, \quad \begin{bmatrix} \overline{u}_{2x}(t) \\ \overline{u}_{2y}(t) \end{bmatrix} = \begin{bmatrix} 0 \\ \text{as given} \end{bmatrix},$$

for node 2, and finally,

$$\begin{bmatrix} u_{3x}(t) \\ u_{3y}(t) \end{bmatrix} = \begin{bmatrix} 0 \\ u_{3y}(t) \end{bmatrix}, \quad \begin{bmatrix} \overline{u}_{3x}(t) \\ \overline{u}_{3y}(t) \end{bmatrix} = \begin{bmatrix} \text{as given} \\ 0 \end{bmatrix},$$

for node 3. In the toolbox code, the free degrees of freedom may be distinguished from the prescribed ones using the attributes of the `field` class. The free degrees of freedom get nonzero equation numbers (attribute `eqnums`), and the magnitudes are stored in the attribute `values`; the prescribed degrees of freedom are marked with the attribute `is_prescribed` and the value to which these degrees of freedom are being set is the `prescribed_value`. The `gather()` method of the `field` class may be used to retrieve all these attributes. A convenient way of looking at the attributes of the `field` object, or at any other SOFEA object for that matter, is the OBgui object browser.

12.3.2 *The test function*

Using the trick described below Eq. (12.12), we will write the test function as

$$\begin{bmatrix} \eta_x(\boldsymbol{x}) \\ \eta_y(\boldsymbol{x}) \\ \eta_z(\boldsymbol{x}) \end{bmatrix} = \sum_{\text{all } i} N_i(\boldsymbol{x}) \begin{bmatrix} \eta_{ix} \\ \eta_{iy} \\ \eta_{iz} \end{bmatrix}, \qquad (12.13)$$

where we define

> $\eta_{ix} = 0$ if the x degree of freedom at node i is prescribed;
> $\eta_{iy} = 0$ if the y degree of freedom at node i is prescribed;
> $\eta_{iz} = 0$ if the z degree of freedom at node i is prescribed;

otherwise $\eta_{ix}, \eta_{iy}, \eta_{iz}$ are arbitrary numbers.

We will also use a more succinct version

$$\begin{bmatrix} \eta_x(\boldsymbol{x}) \\ \eta_y(\boldsymbol{x}) \\ \eta_z(\boldsymbol{x}) \end{bmatrix} = [\eta(\boldsymbol{x})] = \sum_{\text{all } i} N_i(\boldsymbol{x}) \left[\eta_i \right] , \tag{12.14}$$

where $[\eta_i]$ stands for a column matrix holding the components $[\eta_i]_k$ of the vector of the degrees of freedom at node i.

12.3.3   *Producing the requisite equations*

It may be easily verified that the definition of the test function gives us just enough equations to solve for all the free degrees of freedom. Use (12.13) in Eq. (12.10) to obtain

$$\int_V [\boldsymbol{\eta}]^T \rho[\ddot{\boldsymbol{u}}] \, \mathrm{d}V - \int_V [\boldsymbol{\eta}]^T [\overline{\boldsymbol{b}}] \, \mathrm{d}V - \sum_{i=x,y,z} \int_{S_{t,i}} (\boldsymbol{\eta})_i \, \overline{t}_i \, \mathrm{d}S$$

$$+ \int_V (\mathcal{B}[\boldsymbol{\eta}])^T \boldsymbol{D}\mathcal{B}[\boldsymbol{u}] \, \mathrm{d}V = 0 \tag{12.15}$$

$$u_i = \overline{u}_i \quad \text{and} \quad (\boldsymbol{\eta})_i = 0 \quad \text{on } S_{u,i} \quad \text{for } i = x, y, z .$$

The key is to obtain $[\boldsymbol{\eta}]^T$ in all the terms. The trickiest tweaking will be required for $\mathcal{B}[\boldsymbol{\eta}]$. Substituting (12.13), we get

$$\mathcal{B}[\boldsymbol{\eta}] = \mathcal{B} \sum_{\text{all } j} N_j(\boldsymbol{x}) \begin{bmatrix} \eta_{jx} \\ \eta_{jy} \\ \eta_{jz} \end{bmatrix} ,$$

but since the $\eta_{jx}, \eta_{jy}, \eta_{jz}$'s are just numbers, the symmetric gradient operator works with the basis functions as if we simply multiplied the matrix of the operator with a scalar

$$\mathcal{B}[\boldsymbol{\eta}] = \sum_{\text{all } j} \mathcal{B}\left(N_j(\boldsymbol{x})\right) \begin{bmatrix} \eta_{jx} \\ \eta_{jy} \\ \eta_{jz} \end{bmatrix} .$$

Spelled out in full:

$$
\mathcal{B}\left(N_j(\boldsymbol{x})\right) =
\begin{bmatrix}
\dfrac{\partial N_j(\boldsymbol{x})}{\partial x} & 0 & 0 \\[2mm]
0 & \dfrac{\partial N_j(\boldsymbol{x})}{\partial y} & 0 \\[2mm]
0 & 0 & \dfrac{\partial N_j(\boldsymbol{x})}{\partial z} \\[2mm]
\dfrac{\partial N_j(\boldsymbol{x})}{\partial y} & \dfrac{\partial N_j(\boldsymbol{x})}{\partial x} & 0 \\[2mm]
\dfrac{\partial N_j(\boldsymbol{x})}{\partial z} & 0 & \dfrac{\partial N_j(\boldsymbol{x})}{\partial x} \\[2mm]
0 & \dfrac{\partial N_j(\boldsymbol{x})}{\partial z} & \dfrac{\partial N_j(\boldsymbol{x})}{\partial y}
\end{bmatrix} .
$$

Next, the test function is substituted as

$$
\sum_{\text{all } j} [\eta_j]^T \int_V N_j(\boldsymbol{x}) \rho [\ddot{\boldsymbol{u}}] \, \mathrm{d}V - \sum_{\text{all } j} [\eta_j]^T \int_V N_j(\boldsymbol{x})[\overline{\boldsymbol{b}}] \, \mathrm{d}V
$$

$$
- \sum_{i=x,y,z} \sum_{\text{all } j} ([\eta_j])_i \int_{S_{t,i}} N_j(\boldsymbol{x}) \overline{t}_i \, \mathrm{d}S
$$

$$
-+ \sum_{\text{all } j} [\eta_j]^T \int_V \mathcal{B}^T\left(N_j(\boldsymbol{x})\right) \boldsymbol{D} \mathcal{B}[\boldsymbol{u}] \, \mathrm{d}V = 0 \qquad (12.16)
$$

$$
u_i = \overline{u}_i \quad \text{and} \quad (\boldsymbol{\eta})_i = 0 \quad \text{on } S_{u,i} \quad \text{for } i = x,y,z \, .
$$

The components are independent, hence (12.16) should hold for each component separately

$$
\sum_{\text{all } j} [\eta_j]_i \int_V N_j(\boldsymbol{x}) \rho [\ddot{\boldsymbol{u}}]_i \, \mathrm{d}V - \sum_{\text{all } j} [\eta_j]_i \int_V N_j(\boldsymbol{x})[\overline{\boldsymbol{b}}]_i \, \mathrm{d}V
$$

$$
- \sum_{\text{all } j} [\eta_j]_i \int_{S_{t,i}} N_j(\boldsymbol{x}) \overline{t}_i \, \mathrm{d}S
$$

$$
+ \sum_{\text{all } j} [\eta_j]_i \int_V \left[\mathcal{B}^T\left(N_j(\boldsymbol{x})\right) \boldsymbol{D} \mathcal{B}[\boldsymbol{u}] \right]_i \, \mathrm{d}V = 0 \qquad (12.17)
$$

$$
u_i = \overline{u}_i \quad \text{and} \quad (\boldsymbol{\eta})_i = 0 \quad \text{on } S_{u,i} \quad \text{for } i = x,y,z
$$

Regrouping the sums yields finally

$$\sum_{\text{all } j} [\eta_j]_i \left\{ \int_V N_j(\boldsymbol{x})\rho[\ddot{\boldsymbol{u}}]_i \ dV - \int_V N_j(\boldsymbol{x})[\overline{\boldsymbol{b}}]_i \ dV \right.$$

$$\left. - \int_{S_{t,i}} N_j(\boldsymbol{x})\overline{t}_i \ dS + \int_V \left[\boldsymbol{\mathcal{B}}^T \left(N_j(\boldsymbol{x}) \right) \boldsymbol{D}\boldsymbol{\mathcal{B}}[\boldsymbol{u}] \right]_i \ dV \right\} = 0 \qquad (12.18)$$

$$u_i = \overline{u}_i \quad \text{and} \quad (\boldsymbol{\eta})_i = 0 \quad \text{on } S_{u,i} \quad \text{for } i = x, y, z$$

Now we account for some of the $[\eta_j]_i$ components being zero, which happens whenever node j is on the boundary $S_{u,i}$. These components do not field any equations, and the interiors of the braces may be ignored (the $[\eta_j]_i = 0$ make them irrelevant).

The components $[\eta_j]_i$ that are not zero are completely arbitrary. Therefore, the interiors of the braces have to vanish identically, yielding *one equation for each free degree of freedom.*

$$\int_V N_j(\boldsymbol{x})\rho[\ddot{\boldsymbol{u}}]_i \ dV - \int_V N_j(\boldsymbol{x})[\overline{\boldsymbol{b}}]_i \ dV$$

$$- \int_{S_{t,i}} N_j(\boldsymbol{x})\overline{t}_i \ dS + \int_V \left[\boldsymbol{\mathcal{B}}^T \left(N_j(\boldsymbol{x}) \right) \boldsymbol{D}\boldsymbol{\mathcal{B}}[\boldsymbol{u}] \right]_i \ dV = 0 \ , \qquad (12.19)$$

where $u_i = \overline{u}_i$; for all j, and $i = x, y, z$, such that $[\eta_j]_i \neq 0$.

These equations express the **dynamic force equilibrium** at node j in the direction i.

12.4   The discrete equations: system of ODE's

Finally, we substitute the trial function from (12.12). We will use the more streamlined version

$$\begin{bmatrix} u_x(\boldsymbol{x},t) \\ u_y(\boldsymbol{x},t) \\ u_z(\boldsymbol{x},t) \end{bmatrix} = [u(\boldsymbol{x},t)] = \sum_{\text{all } k} N_k(\boldsymbol{x}) \left\{ \left[u_k(t) \right] + \left[\overline{u}_k(t) \right] \right\} \ , \qquad (12.20)$$

where $[u_k]$ stands for a column matrix holding the components $[u_k]_m$ of the vector of the free degrees of freedom at node k; analogously for the prescribed degrees of freedom. To satisfy the essential boundary conditions by interpolation, we set for the prescribed mth component at node k

$$[\overline{u}_k(t)]_m = [\overline{u}(\boldsymbol{x}_k, t)]_m$$

where $\boldsymbol{x}_k$ is the location of node k.

The trial function is substituted into (12.19). To keep things orderly and clear, we will substitute term by term.

12.4.1 Inertial term: Mass matrix

We begin with the inertial effects:

$$\int_V N_j(\boldsymbol{x})\rho[\ddot{\boldsymbol{u}}]_i \, dV =$$

$$\int_V N_j(\boldsymbol{x})\rho \sum_{\text{all } k} N_k(\boldsymbol{x}) \left\{ \left[\ddot{u}_k(t) \right]_i + \left[\ddot{\overline{u}}_k(t) \right]_i \right\} \, dV =$$

$$\sum_{\text{all } k} \int_V N_j(\boldsymbol{x})\rho N_k(\boldsymbol{x}) \, dV \left\{ \left[\ddot{u}_k(t) \right]_i + \left[\ddot{\overline{u}}_k(t) \right]_i \right\} =$$

$$\sum_{\text{all } k} \int_V N_j(\boldsymbol{x})\rho N_k(\boldsymbol{x}) \, dV \, \delta_{im} \left[\ddot{u}_k(t) \right]_m$$

$$+ \sum_{\text{all } k} \int_V N_j(\boldsymbol{x})\rho N_k(\boldsymbol{x}) \, dV \, \delta_{im} \left[\ddot{\overline{u}}_k(t) \right]_m . \tag{12.21}$$

Two terms emerge: firstly, the **inertial force**

$$F_{a,(j,i)} = \sum_{\text{all } k} M_{(j,i)(k,m)} \left[\ddot{u}_k(t) \right]_m , \tag{12.22}$$

produced by the free accelerations which are coupled together by the **consistent mass matrix**

$$M_{(j,i)(k,m)} = \int_V N_j(\boldsymbol{x})\rho N_k(\boldsymbol{x}) \, dV \, \delta_{im} , \tag{12.23}$$

where (j, i) means *equation number* corresponding to component i at node j (which is a free degree of freedom). Note well that $M_{(j,i)(k,m)}$ is a *two-dimensional array*, addressed by the row index (j, i) and the column index (k, m).

Secondly, there is the **inertial load** produced by the acceleration of the supported nodes

$$F_{\overline{a},(j,i)} = \sum_{\text{all } k} \overline{M}_{(j,i)(k,m)} \left[\ddot{\overline{u}}_k(t) \right]_m , \tag{12.24}$$

where the matrix elements $\overline{M}_{(j,i)(k,m)}$ are calculated exactly as those of (12.23), but (k,m) corresponds to component m at node k, which is prescribed.

12.4.2 Body loads and traction loads

The next two terms represent external loads. The **body load vector** component (j,i) corresponding to free component i at node j is

$$F_{b,(j,i)} = \int_V N_j(\boldsymbol{x}) [\overline{\boldsymbol{b}}]_i \, \mathrm{d}V \ . \tag{12.25}$$

The **surface traction load vector** component (j,i) corresponding to free component i at node j is

$$F_{t,(j,i)} = \int_{S_{t,i}} N_j(\boldsymbol{x}) \overline{t}_i \, \mathrm{d}S \ . \tag{12.26}$$

Note well that $F_{\cdot,(j,i)}$ is a *one-dimensional array* (column vector), addressed by the row index (j,i).

12.4.3 Resisting forces: Stiffness matrix

Finally, the trial function (12.20) is substituted into the last term of (12.19).

$$\int_V \left[\mathcal{B}^T \left(N_j(\boldsymbol{x}) \right) \boldsymbol{D}\mathcal{B}[\boldsymbol{u}] \right]_i \, \mathrm{d}V =$$

$$\int_V \left[\mathcal{B}^T \left(N_j(\boldsymbol{x}) \right) \boldsymbol{D}\mathcal{B} \sum_{\text{all } k} N_k(\boldsymbol{x}) \left\{ [u_k(t)] + [\overline{u}_k(t)] \right\} \right]_i \, \mathrm{d}V =$$

$$\sum_{\text{all } k} \int_V \left[\mathcal{B}^T \left(N_j(\boldsymbol{x}) \right) \boldsymbol{D}\mathcal{B} \left(N_k(\boldsymbol{x}) \right) \left\{ [u_k(t)] + [\overline{u}_k(t)] \right\} \right]_i \, \mathrm{d}V =$$

$$\sum_{\text{all } k} \sum_m \left[\int_V \mathcal{B}^T \left(N_j(\boldsymbol{x}) \right) \boldsymbol{D}\mathcal{B} \left(N_k(\boldsymbol{x}) \right) \, \mathrm{d}V \right]_{im} \left\{ [u_k(t)]_m + [\overline{u}_k(t)]_m \right\}$$

$$\tag{12.27}$$

Two contributions result: the first is the **resisting force** produced by the deformed material.

$$F_{r,(j,i)} = \sum_{\text{all } k} K_{(j,i)(k,m)} [u_k(t)]_m \ . \tag{12.28}$$

The matrix generating the resisting force is the **stiffness matrix**

$$K_{(j,i)(k,m)} = \left[\int_V \mathcal{B}^T \left(N_j(\boldsymbol{x}) \right) \boldsymbol{D}\mathcal{B} \left(N_k(\boldsymbol{x}) \right) \, dV \right]_{im}, \qquad (12.29)$$

where (j,i) $[(k,m)]$ means equation number corresponding to component i at node j (component m at node k); both are free degrees of freedom.

The second is the **nonzero-displacement load** due to the deformation induced by prescribed essential boundary conditions. (Remark: In structural analysis, this kind of load is frequently associated with the loading condition called the *support settlement*.)

$$F_{\overline{r},(j,i)} = \sum_{\text{all } k} \overline{K}_{(j,i)(k,m)} \left[\overline{u}_k(t) \right]_m . \qquad (12.30)$$

The elements $\overline{K}_{(j,i)(k,m)}$ are computed exactly as in (12.29), but (k,m) corresponds to component m at node k, which is prescribed.

12.4.4   *Summary of the elastodynamics ODE's*

Summing the forces (12.22), (12.24), (12.25), (12.26), (12.28), and (12.30) yields a system of second-order ordinary differential equations for the free displacements $\left[u_k(t) \right]_m$

$$\sum_{\text{all } k} M_{(j,i)(k,m)} \left[\ddot{u}_k(t) \right]_m + \sum_{\text{all } k} K_{(j,i)(k,m)} \left[u_k(t) \right]_m =$$
$$-F_{\overline{a},(j,i)} - F_{\overline{r},(j,i)} + F_{b,(j,i)} + F_{t,(j,i)} . \qquad (12.31)$$

In the convenient matrix notation, we could write

$$M\ddot{U} + KU = L , \qquad (12.32)$$

where U collects all the free degrees of freedom. These equations could be directly integrated using a Matlab integrator as indicated in Section 3.4, or even more suitably with a specialized mechanical integrator such as the Newmark average-acceleration integrator. Other approaches, such as integration of the harmonic modal equations are often used.

12.5   Constitutive equations of linearly elastic materials

The strain displacement operator (symmetric gradient operator) (11.32) links displacements in terms of their components in the global Cartesian basis to strains. The strains could be expressed in the same Cartesian basis, but need not be. In fact, it will be most useful not to express the strains in the same global Cartesian coordinate system. The motivating factor is the constitutive equation: For some materials it will be important to keep track of the local orientation of the material volume. For instance, fiber reinforced materials will have very different stiffness properties along the fibers as opposed to perpendicularly to the fibers.

The components of the material stiffness matrix (or the material compliance matrix) are to be understood as being expressed in the local coordinate system $e_{\bar{x}}$, $e_{\bar{y}}$, $e_{\bar{z}}$ attached to the material point in the form of (6.53) (except that the transformation matrix has three columns and rows).

12.5.1   *General anisotropic material*

Because of the symmetry of the material stiffness, for the most general elastic material the number of elastic coefficients is only 21 out of the total of 36 elements of the material stiffness matrix: the *general anisotropic material*. Still, to identify all of these constants represents a major experimental effort, and few engineering materials are characterized as fully anisotropic.

12.5.2   *Orthotropic material*

If a material has three mutually orthogonal planes of symmetry, it is known as an *orthotropic material*. For instance wood is often characterized as such type of material, and fiber-reinforced composites are in more sophisticated analyses also treated as orthotropic. The compliance matrix

$$C = D^{-1} \, ,$$

has a pleasingly simple appearance

$$
C = \begin{bmatrix}
E_1^{-1}, & -\dfrac{\nu_{12}}{E_1}, & -\dfrac{\nu_{13}}{E_1}, & 0, & 0, & 0 \\[2mm]
-\dfrac{\nu_{12}}{E_1}, & E_2^{-1}, & -\dfrac{\nu_{23}}{E_2}, & 0, & 0, & 0 \\[2mm]
-\dfrac{\nu_{13}}{E_1}, & -\dfrac{\nu_{23}}{E_2}, & E_3^{-1}, & 0, & 0, & 0 \\[2mm]
0, & 0, & 0, & G_{12}^{-1}, & 0, & 0 \\[2mm]
0, & 0, & 0, & 0, & G_{13}^{-1}, & 0 \\[2mm]
0, & 0, & 0, & 0, & 0, & G_{23}^{-1}
\end{bmatrix}.
$$

All nine coefficients are independent, and need to be provided as input [Hyer (1998)]. The material stiffness matrix is a bit of a mess. The nonzero elements are

$$
D_{11} = \frac{C_{22}C_{33} - C_{23}C_{23}}{C}, \qquad D_{12} = \frac{C_{13}C_{23} - C_{12}C_{33}}{C},
$$
$$
D_{13} = \frac{C_{12}C_{23} - C_{13}C_{22}}{C}, \qquad D_{22} = \frac{C_{33}C_{11} - C_{13}C_{13}}{C},
$$
$$
D_{23} = \frac{C_{12}C_{13} - C_{23}C_{11}}{C}, \qquad D_{33} = \frac{C_{11}C_{22} - C_{12}C_{12}}{C},
$$
$$
D_{44} = G_{12}, \qquad\qquad D_{55} = G_{13}, \qquad\qquad D_{66} = G_{23}.
$$

Here $C = C_{11}C_{22}C_{33} - C_{11}C_{23}C_{23} - C_{22}C_{13}C_{13} - C_{33}C_{12}C_{12} + 2C_{12}C_{23}C_{13}$.

12.5.3 *Transversely isotropic material*

If a material has an infinite number of planes of symmetry passing through an axis (in this case, the local x-axis), and one plane of symmetry perpendicular to this axis, it is known as a ***transversely isotropic material***. Unidirectionally reinforced composites are of this type, as are for instance muscles. The direction of the fibers is special (oriented along the local x-axis), but the material behaves isotropically in the planes perpendicular to the fibers. Layered materials are also modeled as transversely isotropic: here the direction perpendicular to the layers is special. Figure 12.2 offers an illustration of these two types.

The compliance matrix is obtained from the orthotropic compliance by setting $E_2 = E_3$, $\nu_{12} = \nu_{13}$, $G_{12} = G_{13}$, and importantly

$$
G_{23} = \frac{E_2}{2(1 + \nu_{23})},
$$

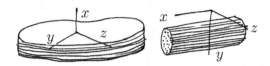

Fig. 12.2 Transversely isotropic model is appropriate for layered or fiber-reinforced materials.

requiring five independent constants, E_1, E_2, ν_{12}, G_{12}, and ν_{23}.

12.5.4   *Isotropic material*

If a material has an infinite number of planes of symmetry of all possible orientations, it is known as an ***isotropic material***. The compliance matrix of isotropic material is based on two material properties, for instance the Young's modulus E and Poisson's ratio ν

$$
C = \begin{bmatrix}
E^{-1}, & -\dfrac{\nu}{E}, & -\dfrac{\nu}{E}, & 0, & 0, & 0 \\[2mm]
-\dfrac{\nu}{E}, & E^{-1}, & -\dfrac{\nu}{E}, & 0, & 0, & 0 \\[2mm]
-\dfrac{\nu}{E}, & -\dfrac{\nu}{E}, & E^{-1}, & 0, & 0, & 0 \\[2mm]
0, & 0, & 0, & G^{-1}, & 0, & 0 \\[2mm]
0, & 0, & 0, & 0, & G^{-1}, & 0 \\[2mm]
0, & 0, & 0, & 0, & 0, & G^{-1}
\end{bmatrix}.
$$

The shear modulus G is not independent, but is expressed as

$$
G = \frac{E}{2(1+\nu)} .
$$

The material stiffness is then

$$
D = \begin{bmatrix}
\lambda + 2G & \lambda & \lambda & 0 & 0 & 0 \\
\lambda & \lambda + 2G & \lambda & 0 & 0 & 0 \\
\lambda & \lambda & \lambda + 2G & 0 & 0 & 0 \\
0 & 0 & 0 & G & 0 & 0 \\
0 & 0 & 0 & 0 & G & 0 \\
0 & 0 & 0 & 0 & 0 & G
\end{bmatrix} ,
$$

where we introduce the Lamé constant

$$\lambda = \frac{E\nu}{(1+\nu)(1-2\nu)}$$

for convenience.

12.6 Imposed (thermal) strains

Often the material from which a structure is built up reacts to the environment by deformation. The material experiences the environment in possibly different ways or different measures in different locations, and stresses are produced.

To get started, think about a very small piece of material that is exposed to the environment so that we can assume that the resultant relative deformation is homogeneous and no stress is produced. For the sake of this argument, let us consider one particular environmental effect: *thermal expansion*. However, similar effects may be produced by shrinkage, swelling, piezoelectric effects, and so on.

When the small sample of material is at a *reference temperature* it is unstressed, and we define its displacements and the associated strains to be zero in this state: the *reference state*. Then the temperature is increased by ΔT, and the material responds by displacement, and because by assumption the deformation is homogeneous, the entire sample experiences uniform strains. Based on experimental evidence, the following model is adopted for orthotropic materials to describe the strains in coordinates aligned with the material directions

$$[\epsilon_\Theta] = \begin{bmatrix} \epsilon_x \\ \epsilon_y \\ \epsilon_z \\ \gamma_{xy} \\ \gamma_{xz} \\ \gamma_{yz} \end{bmatrix} = \Delta T \begin{bmatrix} \alpha_x \\ \alpha_y \\ \alpha_z \\ 0 \\ 0 \\ 0 \end{bmatrix}. \tag{12.33}$$

Evidently, the thermal expansion is assumed not to cause any shear strains. The factors $\alpha_x, \alpha_y, \alpha_z$ are the so-called *coefficients of thermal expansion*; different in different directions, in general.

In addition to the imposed strain ϵ_Θ, the sample is also exposed to stresses on its boundary, again such that they result in uniform strain. The total strains that the sample experiences consist of the imposed strains to which the mechanical strains are added. Since the mechanical strains are available from the stresses through the constitutive equation, we write

$$\epsilon = \epsilon_{\text{total}} = \epsilon_\Theta + C\sigma \ . \tag{12.34}$$

Therefore, we have a modification of the constitutive equation

$$\sigma = D \left(\epsilon - \epsilon_\Theta \right) \ . \tag{12.35}$$

The total strain is related to the displacement via the strain-displacement relation (11.32), and thus we may write

$$\sigma = D \left(\mathcal{B}u - \epsilon_\Theta \right) \ .$$

This is sometimes also presented as

$$\sigma = D\mathcal{B}u + \sigma_\Theta \ ,$$

where we introduce the so-called **thermal stress** σ_Θ, but purely as a convenience; the primary quantity is the measurable thermal strain.

As expected, the residual equation form that enters the discretization process needs to be augmented with respect to (12.10) to become

$$\int_V \boldsymbol{\eta} \cdot \rho \ddot{\boldsymbol{u}} \ dV - \int_V \boldsymbol{\eta} \cdot \overline{\boldsymbol{b}} \ dV - \sum_{i=x,y,z} \int_{S_{t,i}} (\boldsymbol{\eta})_i \, \overline{t}_i \ dS \tag{12.36}$$

$$+ \int_V (\mathcal{B}\boldsymbol{\eta}) \cdot D \left(\mathcal{B}\boldsymbol{u} - \epsilon_\Theta \right) \ dV = 0 \tag{12.37}$$

$$u_i = \overline{u}_i \quad \text{and} \quad (\boldsymbol{\eta})_i = 0 \quad \text{on } S_{u,i} \quad \text{for } i = x, y, z$$

Clearly, there will be one more term to discretize

$$- \int_V (\mathcal{B}\boldsymbol{\eta}) \cdot D\epsilon_\Theta \ dV \ , \tag{12.38}$$

yielding the **thermal strain load**

$$F_{\Theta,(j,i)} = \left[\int_V \mathcal{B}^T \left(N_j(\boldsymbol{x}) \right) D\epsilon_\Theta \ dV \right]_i \ . \tag{12.39}$$

Therefore, with the inclusion of the thermal strains, the system of ordinary differential equations (12.31) that describe the discrete problem becomes

$$\sum_{\text{all } k} M_{(j,i)(k,m)} \left[\ddot{u}_k(t) \right]_m + \sum_{\text{all } k} K_{(j,i)(k,m)} \left[u_k(t) \right]_m =$$

$$- \sum_{\text{all } k} \overline{M}_{(j,i)(k,i)} \left[\ddot{\overline{u}}_k(t) \right]_i - \sum_{\text{all } k} \overline{K}_{(j,i)(k,m)} \left[\overline{u}_k(t) \right]_m$$

$$+ F_{b,(j,i)} + F_{t,(j,i)} + F_{\Theta,(j,i)} . \qquad (12.40)$$

12.7 Strain-displacement matrix

The elements of the stiffness matrix (12.29) are evaluated using numerical quadrature element-by-element (explained in detail in Section 6.8 for the conductivity matrix). This operation produces element-level stiffness matrices that couple together the finite element nodes of each element. To simplify the notation, we will define the strain-displacement matrix as applied to a single basis function (the nodal strain-displacement matrix)

$$\boldsymbol{B}_k^e = \mathcal{B}\left(N_k(\boldsymbol{x}) \right) \quad \text{for } \boldsymbol{x} \in \text{element } e , \qquad (12.41)$$

where k indicates the node number, and e identifies the element (it bears emphasis that the strain-displacement matrix of node k is different in each element that shares this node).

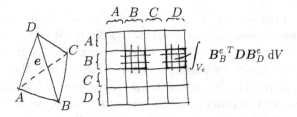

Fig. 12.3   Assembly of the element stiffness matrix.

Consider now a structure that consists of a single element, say a tetrahedron with four nodes, A, B, C and D (Fig. 12.3). Let us also assume that all the degrees of freedom are free (hypothetically: if they really were, the stiffness matrix would be singular). The stiffness matrix of such a structure (i. e. of this single element) has 12 rows and columns.

$$K_{(j,i)(k,m)} = \left[\int_{V_e} \boldsymbol{B}_j^{eT} \boldsymbol{D} \boldsymbol{B}_k^e \, \mathrm{d}V \right]_{im} , \quad j,k = A, B, C, D . \qquad (12.42)$$

For j, k fixed, say $j = B$ and $k = D$, the matrix

$$\int_{V_e} \boldsymbol{B}_B^{e\,T} \boldsymbol{D} \boldsymbol{B}_D^e \, \mathrm{d}V \, , \qquad (12.43)$$

is a 3×3 submatrix of the element stiffness matrix: compare with the illustration in Fig. 12.3 (the off-diagonal block). Similarly, for $j = B$ and $k = B$ we obtain the diagonal block. Therefore, the entire element stiffness matrix may be computed in one shot as

$$\boldsymbol{K}_e = \int_{V_e} \boldsymbol{B}^{e\,T} \boldsymbol{D} \boldsymbol{B}^e \, \mathrm{d}V \, , \qquad (12.44)$$

using the blocked matrix (the element strain-displacement matrix)

$$\boldsymbol{B}_e = [\boldsymbol{B}_A^e, \boldsymbol{B}_B^e, \boldsymbol{B}_C^e, \boldsymbol{B}_D^e] \, . \qquad (12.45)$$

Correspondingly, the displacements at the nodes will be ordered into a column vector of displacement components in the global Cartesian basis (with the Matlab syntax: semicolon means new line)

$$\boldsymbol{U}_e = [[u_A]; [u_B]; [u_C]; [u_D]] \, . \qquad (12.46)$$

12.7.1 *Transformation of basis*

It remains to discuss the issue of the choice of coordinate systems when evaluating the stiffness matrix integrals. As discussed in Section 12.5, the components of the material stiffness matrix are expressed on the local Cartesian basis of material orientation directions. Referring to the definition of the stiffness matrix (12.29), we see that there is a need to transform between the local material directions and the global Cartesian basis. Drawing on the example of the element stiffness matrix (12.44), the element's restoring force may be expressed as

$$\boldsymbol{F}_e = \boldsymbol{K}_e \boldsymbol{U}_e = \int_{V_e} \boldsymbol{B}^{e\,T} \boldsymbol{D} \boldsymbol{B}^e \, \mathrm{d}V \, \boldsymbol{U}_e \, , \qquad (12.47)$$

While the displacement components in $\boldsymbol{U}_e$ and the force components in $\boldsymbol{F}_e$ are in the global Cartesian basis, the material stiffness matrix is expressed on the basis of the local material directions. Evidently, the linear algebra operations between the material matrix and displacements/forces must incorporate the transformation from one basis to another.

There are several ways in which this transformation could be carried out. In the **SOFEA** toolbox the following approach is used: the strain-displacement matrix is defined to produce strains in the local material basis, while taking displacement components in the global basis. This can be accomplished by writing

$$[\epsilon]^{(\overline{x})} = \mathcal{B}^{\overline{x}}[\boldsymbol{u}]^{(\overline{x})} = \mathcal{B}^{(\overline{x})}\left([T][\boldsymbol{u}]^{(x)}\right) =$$
$$\left(\mathcal{B}^{(\overline{x})}[T]\right)[\boldsymbol{u}]^{(x)} = \mathcal{B}^{(\overline{x},x)}[\boldsymbol{u}]^{(x)} \qquad (12.48)$$

where we use the superscript $(\overline{x})$ or (x) to indicate in which coordinate system the components are expressed. The strain-displacement operator $\mathcal{B}^{(\overline{x},x)} = \mathcal{B}^{(\overline{x})}[T]$ incorporates the geometric transformation $[T]$ of the displacement vector components from the global basis into the local material directions basis. This transformation may be derived from the equality

$$\left[\boldsymbol{e}_x, \boldsymbol{e}_y, \boldsymbol{e}_z\right] \begin{bmatrix} u_x \\ u_y \\ u_z \end{bmatrix} = \left[\boldsymbol{e}_{\overline{x}}, \boldsymbol{e}_{\overline{y}}, \boldsymbol{e}_{\overline{z}}\right] \begin{bmatrix} u_{\overline{x}} \\ u_{\overline{y}} \\ u_{\overline{z}} \end{bmatrix}. \qquad (12.49)$$

Pre-multiplying with

$$\begin{bmatrix} \boldsymbol{e}_{\overline{x}} \\ \boldsymbol{e}_{\overline{y}} \\ \boldsymbol{e}_{\overline{z}} \end{bmatrix}$$

leads to the transformation of vector components

$$[\boldsymbol{R}_m]^T \begin{bmatrix} u_x \\ u_y \\ u_z \end{bmatrix} = [T] \begin{bmatrix} u_x \\ u_y \\ u_z \end{bmatrix} = \begin{bmatrix} u_{\overline{x}} \\ u_{\overline{y}} \\ u_{\overline{z}} \end{bmatrix}. \qquad (12.50)$$

The transformation $[T]$ is thus identified with an orthogonal (rotation) matrix

$$[\boldsymbol{R}_m] = \begin{bmatrix} \boldsymbol{e}_{\overline{x}} \cdot \boldsymbol{e}_x, & \boldsymbol{e}_{\overline{y}} \cdot \boldsymbol{e}_x, & \boldsymbol{e}_{\overline{z}} \cdot \boldsymbol{e}_x \\ \boldsymbol{e}_{\overline{x}} \cdot \boldsymbol{e}_y, & \boldsymbol{e}_{\overline{y}} \cdot \boldsymbol{e}_y, & \boldsymbol{e}_{\overline{z}} \cdot \boldsymbol{e}_y \\ \boldsymbol{e}_{\overline{x}} \cdot \boldsymbol{e}_z, & \boldsymbol{e}_{\overline{y}} \cdot \boldsymbol{e}_z, & \boldsymbol{e}_{\overline{z}} \cdot \boldsymbol{e}_z \end{bmatrix}, \qquad (12.51)$$

which is just a three-dimensional analog of the two-dimensional transformation (6.53).

The global-to-local ***strain-displacement matrix*** is therefore defined as

$$\boldsymbol{B}_k^e = \mathcal{B}^{(\overline{x},x)}\left(N_k(\boldsymbol{x})\right) = \mathcal{B}^{(\overline{x})}\left(N_k(\boldsymbol{x})\right)\left[\boldsymbol{R}_m\right]^T \text{ for } \boldsymbol{x} \in \text{element } e . \quad (12.52)$$

The $\boldsymbol{B}_k^e$ nodal matrices are used to compose the element strain-displacement matrix (12.45). In the SOFEA toolbox, the strain-displacement matrix is computed for a three-dimensional solid geometric cell by the method Blmat:

```
0023 function B = Blmat¹(self, pc, x, Rm)
0024     Nder = bfundpar (self, pc);
0025     Ndersp=bfundsp(self,Nder,x*Rm);
0026     nfens = size(Ndersp, 1);
0027     dim=3;
0028     B = zeros(6,nfens*dim);
0029     for i= 1:nfens
0030         B(:,dim*(i-1)+1:dim*i)...
0031         = [ Ndersp(i,1) 0           0           ; ...
0032             0           Ndersp(i,2) 0           ; ...
0033             0           0           Ndersp(i,3) ; ...
0034             Ndersp(i,2) Ndersp(i,1) 0           ; ...
0035             Ndersp(i,3) 0           Ndersp(i,1) ; ...
0036             0           Ndersp(i,3) Ndersp(i,2) ]*Rm';
0037     end
0038     return;
0039 end
```

The input arguments are the parametric coordinates pc, the array of nodal coordinates x (the 3-D version of (6.42)), and the transformation matrix $\boldsymbol{R}_m$. Comparing with the definitions (12.52) and (12.45), the Matlab code is an almost literal translation of these formulas. Note that the global location vectors for each node x are transformed into the local Cartesian basis of material directions, x*Rm, as the matrix $\mathcal{B}^{(\overline{x})}\left(N_k(\boldsymbol{x})\right)$ in Eq. (12.52) requires derivatives with respect to the material basis.

The method Blmat is for solid elements defined for the common ancestor class of all solid elements, gcell_3_manifold class. Analogously, one-dimensional and two-dimensional elements have their versions defined by the gcell_1_manifold and gcell_2_manifold classes; in other words,

---

[1] Folder: SOFEA/classes/gcell/@gcell_3_manifold

the method `Blmat` is not specific to any particular element type or shape.

12.8   Stiffness matrix

In Chapter 3 and further in Section 6.8, it was pointed out that all the problem dependent code was kept in a descendent of the `feblock` class. The elastodynamics model of this chapter is implemented in the `feblock_defor_ss` finite element block class.

The stiffness matrix (12.29) is assembled from the element-wise stiffness matrices (12.44). These are calculated by the `stiffness` method, and returned in the `ems` array. The `stiffness` method takes as arguments the geometry of the mesh (field `geom`), in order to have access to the locations of the nodes, and the displacement field u to provide the global equation numbers of the unknowns.

```
0008 function ems = stiffness² (self, geom, u)
0009     gcells = get(self.feblock,'gcells');
0010     nfens = get(gcells(1),'nfens');
0011     ems(1:length(gcells)) = deal(elemat);% Pre-allocate
0012     % Integration rule
0013     integration_rule=get(self.feblock,'integration_rule');
0014     pc = get(integration_rule, 'param_coords');
0015     w  = get(integration_rule, 'weights');
0016     npts_per_gcell = get(integration_rule, 'npts');
0017     % Material
0018     mat = get(self.feblock, 'mater');
```

The material stiffness matrix is pre-allocated, and the loop over all the geometric cells within the block begins. The connectivity of the geometric cell is retrieved, and, based on the connectivity, the locations of the nodes x are gathered from the geometry field `geom`. Finally, the matrix `Ke` is zeroed out in preparation for the numerical integration.

```
0019     Ke = zeros(get(geom,'dim')*nfens); % preallocate
0020     % Now loop over all gcells in the block
0021     for i=1:length(gcells)
0022         conn = get(gcells(i), 'conn'); % connectivity
0023         x = gather(geom, conn, 'values', 'noreshape'); % coord
```

---

[2]Folder: `SOFEA/classes/feblock/@feblock_defor_ss`

```
0024          Ke = 0*Ke; % zero out element stiffness
```

The loop over the quadrature points begins by calculating the basis functions in order to be able to evaluate the location of the current integration point.

```
0025          % Loop over all integration points
0026          for j=1:npts_per_gcell
0027              N = bfun(gcells(i),pc(j,:));
```

The matrix of the local material directions (material orientation matrix) is computed. It may depend on the location of the quadrature point, and/or on the tangents to the parametric curves passing through the quadrature point, or indeed on any other user-defined parameter. If required, these quantities would be calculated by the method material_directions from the supplied arguments.

```
0028          Rm = material_directions(self,
                      gcells(i),pc(j,:),x);
```

Compute the Jacobian associated with the integration point, and since the integration is to be performed over the 3-D volume, the method Jacobian_volume needs to be used; for the solid elements it is identical to the Jacobian method.

```
0029          detJ = Jacobian_volume(gcells(i),pc(j,:),x);
```

The strain-displacement matrix for the element is calculated from the spatial gradients of the basis functions, and the matrix of material directions.

```
0030          B = Blmat(gcells(i), pc(j,:), x, Rm);
```

The material stiffness matrix is calculated by the material object mat, and since it may change from point to point, the location of the current integration point, N'*x, is passed to the method.

```
0031          D = tangent_moduli(mat,struct('xyz',N'*x));
```

The product of the strain-displacement matrix and the tangent moduli (the material stiffness matrix) is accumulated in the element stiffness matrix Ke.

```
0032          Ke = Ke + B'*D*B * detJ * w(j);
0033      end
```

Finally, the computed element stiffness matrix is stored in the `ems(i)` object of the `elemat` class. Note that the equation numbers are gathered from the displacement field.

```
0034            ems(i) = set(ems(i), 'mat', Ke);
0035            ems(i) = set(ems(i), 'eqnums',
                                gather(u,conn,'eqnums'));
0036     end
0037     return;
0038 end
```

12.9 Pure-traction problems and singular stiffness

For pure-traction problems, first introduced in Section 11.6.6, the stiffness matrix will become singular (not of full rank). It doesn't take too much effort to verify that strains computed from the rigid-body displacements (11.43) are all identically zero. Therefore, also the deformation energy induced by the rigid body motion is zero, as is easily verified by referring to (11.36). Another way of calculating the energy of deformation is by invoking the formula for the strains in terms of the element displacements, $\epsilon = B^e U_e$, and the assembly of the element stiffness matrices:

$$\Phi(\epsilon) = \int_V \phi(\epsilon)\,dV = \sum_e \int_{V_e} \phi(\epsilon)\,dV =$$

$$\int_V \frac{1}{2}\epsilon^T D\epsilon\,dV = \sum_e \frac{1}{2}U_e^T \int_{V_e} B^{eT} DB^e\,dV\,U_e =$$

$$\sum_e \frac{1}{2}U_e^T K_e U_e = \frac{1}{2}U^T KU . \tag{12.53}$$

Since this energy vanishes for a nonzero displacement $U \neq 0$, the global matrix K must be singular to produce a positive semi-definite quadratic form $\frac{1}{2}U^T KU \geq 0$. The stiffness matrix falls short of the full rank by the number of possible rigid body modes: six, when all three rotations and translations are possible, or less. The rigid-body displacements must be prevented; additional supports, or springs grounding the structure are common solutions.

Exercises

(1) For an isotropic material, consider whether the material parameter values produce a reasonable (that is positive definite) material stiffness matrix. Provide sufficient detail to argue your point.

   (a) Young's modulus $E = 0$, Poisson's ratio $\nu > 0$.
   (b) Young's modulus $E > 0$, Poisson's ratio $\nu = 0$.
   (c) Young's modulus $E > 0$, Poisson's ratio $\nu = 1/2$.
   (d) Young's modulus $E > 0$, Poisson's ratio $\nu = -1/4$.

Chapter 13

Finite Elements for True 3-D Problems

With the formulation sketched out, it only remains to pick a finite element to be able to perform an analysis. We begin with the tetrahedron T4. All the necessary formulas have been put into place in Section 9.5, and we may immediately proceed to the first example.

The equation of motion (12.32) stands for the so-called free vibration when there is no forcing, $L = 0$; compare also with (3.5). The built-in Matlab eigenvalue solver will be used to obtain the numerical solution, and the task for the SOFEA script is relatively simple: compute the stiffness matrix and the mass matrix.

13.1 Modal analysis with the tetrahedron T4: the drum

As our first example, we consider the vibration of a moderately thick circular plate as shown in Fig. 13.1. The cylindrical surface is fully clamped (all displacements zero). The material is isotropic. The analytical solution has been worked out in dependence on the number of "nodes" (locations of approximately zero displacement) radially and circumferentially [Blevins (2001)]. Therefore, this is a very good example with which to test the finite element formulation.

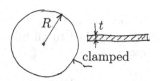

Fig. 13.1   Clamped circular plate (drum).

The Matlab script `drum_t4`[1] solves the vibration problem for the four lowest eigenvalues (shown in Fig. 13.2). First, a few variables are defined and a mesh is produced using the mesh function `t4cylinderdel`. The mesh is relatively coarse, with only two element edges through the thickness.

```
0001 E=0.1e6;% Pa
0002 nu=0.3;
0003 rho=1000;% kg
0004 R= 25.0e-3;% m
0005 t= 2.0e-3;% m
0006 rand('state',0);% try to comment out this line and compare
0007 %                      results for several subsequent runs
0008
0009 % Mesh
0010 [fens,gcells] = t4cylinderdel(t,R, 2,7);
```

The material is small-strain (`ss`), linearly elastic (`linel`), isotropic (`iso`), and triaxial (`triax`). The property class `property_linel_iso` supplies methods to compute the material stiffness matrix for this type of material. Each type of material has its own property class, and the material class `mater_defor_ss_linel_triax` undertakes to insulate the methods of the finite element block from the details of the properties. Note that the mass density needs also be supplied (dynamics!).

```
0012 prop=property_linel_iso(struct('E',E,'nu',nu,'rho',rho));
0013 mater=mater_defor_ss_linel_triax(struct('property',prop));
```

The finite element block is of class `feblock_defor_ss`. The integration rule is a one-point quadrature, provided by the class `tet_rule`.

```
0013 % Finite element block
0014 feb = feblock_defor_ss (struct ('mater',mater,
         'gcells',gcells,...
0015     'integration_rule',tet_rule (1)));
```

The geometry field and the displacement field are set up as usual. The essential boundary condition is applied to all finite element nodes on the cylindrical surface (line 0024).

```
0018 geom=field(struct ('name',['geom'],'dim',3,'fens',fens));
0019 % Define the displacement field
```

---

[1]Folder: `SOFEA/examples/stress`

```
0020 u    = 0*geom; % zero out
0021 % Apply EBC's
0022 for i=1:length(fens)
0023     xyz = get(fens(i),'xyz');
0024     if abs(R-norm(xyz(2:3))) <0.001*R
0025         u    = set_ebc(u, i, [1], [], 0.0);
0026     end
0027 end
0028 u    = apply_ebc (u);
0029 % Number equations
0030 u    = numbereqns (u);
```

The two methods, **stiffness** and **mass**, discussed previously are invoked. Note that the method **mass** computes the consistent element mass matrices, and the free vibration problem solution will be sought with a consistent mass matrix.

```
0032 K = start (sparse_sysmat, get(u, 'neqns'));
0033 K = assemble (K, stiffness(feb, geom, u));
0034 M = start (sparse_sysmat, get(u, 'neqns'));
0035 M = assemble (M, mass(feb, geom, u));
```

Finally, the built-in Matlab solver **eigs** is called. Four lowest eigenvalues are requested (the flag 'SM'). The returned eigenvalues need to be sorted from the smallest to the largest (line 0039-0040).

```
0037 neigvs = 4;
0038 [W,Omega]=eigs(get(K,'mat'),get(M,'mat'),neigvs,'SM');
0039 [Omegas,ix]=sort(diag(Omega));
0040 Omega= diag(Omegas);
```

The plotting section of the script is omitted, but the resulting shapes and the calculated frequencies are summarized in Fig. 13.2: taking the analytical solution as reference, it is clear that the errors are huge. Evidently, a much more refined mesh would be required to obtain reasonable answers (recall: smaller elements, smaller error). A valuable observation: all the calculated values are above the reference frequencies. This is called "convergence from above", and it is an indication that the discrete model (in other words, the T4 finite element) is *too stiff*.

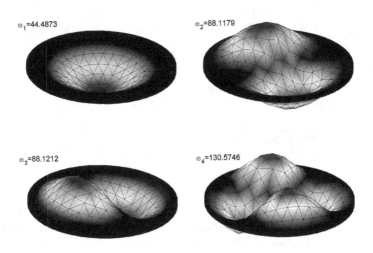

Fig. 13.2   Mode shapes of thick clamped plate. The analytical solution for the natural frequencies yields for these modes $\omega_1 = 15.7511$Hz, $\omega_2 = \omega_3 = 32.7659$Hz, $\omega_4 = 53.757$Hz.

13.2   Modal analysis with the tetrahedron T4: the composite rod

As the second example we will again consider a free vibration problem, but this time the material of the structure will be modeled as transversely isotropic. The structure is a straight rod of circular cross-section, manufactured from carbon fiber infused with polymer resin (Fig. 13.3). The reinforcing fibers are twisted, and for lack of other information, we assume that the twist angle decreases from the outside surface ($15°$) to zero along its axis. The rod is clamped at both ends. Of interest is the lowest natural frequency of the structure.

Fig. 13.3   The composite rod.

The Matlab solution of the problem is in the form of a function, twist_t4[2], in order to allow for a convergence analysis to be performed on a series of meshes. Therefore, an internal function will be run repeatedly to solve for the first natural frequency for different resolutions radially (number of element edges nR) and longitudinally (number of element edges nt).

```
0001 function twist_t4
0002     nR = [2, 3, 4];
0003     nt = [30, 40, 50, 60, 70, 80];
0004     fs =zeros(length(nR),length(nt));
0005     for i=1:length(nR)
0006         for j=1:length(nt)
0007             fs(i,j) =do_twist_t4(nR(i),nt(j))
0008         end
0009         save fs
0010     end
0011 end
0012
0013
0014 function frequency1 =do_twist_t4(nR,nt)
0015 ...
```

The transversely isotropic material from Section 12.5.3 requires the definition of the local material orientation matrix (12.51) at each integration point. The axis $\overline{x}$ needs to be oriented along the fibers; the orientation of the remaining two basis vectors in the isotropy plane is arbitrary. In the function twist, the orientation matrix is derived by first turning a basis triad around the first vector (line 0035-0037), and then twisting the intermediate triad by an angle ramped up from 0° to 15° proportionally to the distance from the axis of the rod (line 0038).

```
0031     function Rm = twist (XYZ, ts)
0032         r= norm(XYZ( 2:3));
0033         if r>0
0034             y=XYZ(2);z=XYZ(3);
0035             e2 = [0, y/r, z/r]';
0036             e3 = skewmat([1, 0, 0])*e2;
0037             Rm= [[1, 0, 0]',e2,e3];
```

---

[2]Folder: SOFEA/examples/stress

```
0038                    Rm=rotmat(r/R*twist_angle*skewmat(e2))*Rm;
0039          else
0040              Rm= eye(3);
0041          end
0042     end
```

The property class **property_linel_transv_iso** defines the material properties for the transversely isotropic material model. Note that five independent material constants need to be supplied.

```
0046     prop = property_linel_transv_iso (...
0047          struct('E1',E1, 'E2',E2,'G12',G12,
                    'nu12',nu12,'nu23',nu23,'rho',rho));
0048     mater = mater_defor_ss_linel_triax (
                    struct('property',prop));
```

It is noteworthy that the function **twist** that defines the directions of the local material basis is supplied to the finite element block as a function handle. Call for Section 12.8 for details on the use of the material orientation matrix.

```
0050     feb = feblock_defor_ss (struct ('mater',mater,
                    'gcells',gcells,...
0051          'integration_rule',tet_rule (integration_order),
                    'Rm',@twist));
```

The rest of the function **twist_t4** is omitted. The first mode shape of the composite rod for a relatively fine mesh is shown in Fig. 13.4: notice that the rod twists as well as bends due to the orientation of the reinforcing fibers in the form of a helix.

Let us now address the issue of convergence: The lowest frequency has been computed for a number of meshes with varying number of elements radially, and along the length of the rod. The lowest natural frequency is displayed for these different meshes in Fig. 13.5. The frequency varies from approximately 2300Hz to slightly less than 1900Hz. It should be noted that the first natural frequency decreases in magnitude with refinement (convergence from above), but the resulting surface is not smooth. (The bad shapes of some elements in the mesh have a profound effect on the accuracy.) The numerical answers are still changing significantly with the refinement, and it is therefore not clear how far from the "exact" solution we might be.

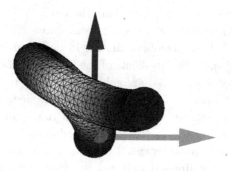

Fig. 13.4   The shape of the first eigenmode.

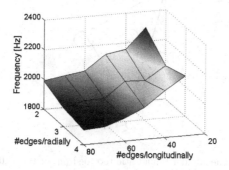

Fig. 13.5   Convergence of the lowest natural frequency with the T4 element.

13.3   Tetrahedron T10

The quadratic tetrahedron is a simple extension of the quadratic triangle to three dimensions. The same idea of constructing the basis functions as (normalized) products of planes will work, yielding for the basis functions the quadratic expressions in the parametric coordinates ξ, η, ζ

$$
\begin{aligned}
N_1 &= (1 - \xi - \eta - \zeta)(2(1 - \xi - \eta - \zeta) - 1) \,, \quad N_2 = \xi(2\xi - 1) \,, \\
N_3 &= \eta(2\eta - 1) \,, \quad N_4 = \zeta(2\zeta - 1) \,, \quad N_5 = 4(1 - \xi - \eta - \zeta)\xi \,, \\
N_6 &= 4\xi\eta \,, \quad N_7 = 4\eta(1 - \xi - \eta - \zeta) \,, \quad N_8 = 4(1 - \xi - \eta - \zeta)\zeta \,, \\
&\qquad\qquad N_9 = 4\xi\zeta \,, \quad N_{10} = 4\eta\zeta \,.
\end{aligned} \tag{13.1}
$$

Since the basis functions are quadratic in the parametric coordinates, their gradients with respect to the parametric coordinates will be linear functions of ξ, η, ζ. Provided the Jacobian matrix in equation (6.40) is constant (independent of ξ, η, ζ), the gradients of the basis functions with respect to x, y, z as computed from (6.35) are going to be linear in those coordinates (see Section 10.6 for details). Hence, computing the elements of the stiffness matrix can be done exactly with the four-point rule from Table 9.2. To the contrary, for curved elements (when a mid-edge node is moved from the mean of the locations of the corners) the four-point rule is not going to be able to integrate the stiffness matrix exactly. However, when the distortion of the element is not excessive, the integration error is not significant.

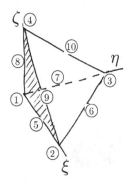

Fig. 13.6   The standard quadratic tetrahedron for the T10 element.

13.3.1   *Example: the drum revisited*

The Matlab script `drum_t10`[3] is a variation on `drum_t4` which applies the quadratic element instead of T4. The natural frequencies obtained for different meshes with the two elements, T4, and T10, are illustrated in Fig. 13.7. While increasing the number of unknowns by making the elements (uniformly) smaller leads generally to more accurate answers, the element T4 is clearly not performing very well. Even for a fairly fine mesh, the results are probably not of much use. On the contrary, the element T10 produces estimates of the frequencies which are of good engineering accuracy (within a fraction of a percent). The first frequency is estimated quite well even with the coarsest mesh (only a single element through the

---

[3] Folder: `SOFEA/examples/stress/3-D`

thickness, and two elements radially).

The graph of Fig. 13.7 may serve as a crude guide to the relative accuracy of the two tetrahedral elements. Sometimes the linear tetrahedron will fare better than in this example, oftentimes much worse.

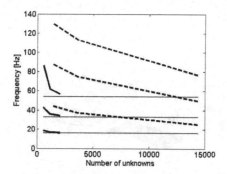

Fig. 13.7   Comparison of the first four natural frequencies of the drum computed with the two tetrahedral elements, the T4 (dashed line) and the T10 (solid line). The analytical solution for the natural frequencies, $\omega_1 = 15.7511$Hz, $\omega_2 = \omega_3 = 32.7659$Hz, $\omega_4 = 53.757$Hz, is indicated with horizontal lines.

13.4   The composite rod with the tetrahedron T10

The problem from Section 13.2 is addressed here with the quadratic tetrahedron T10. Remarkably, the Matlab function twist_t10[4] differs from twist_t4 in only two lines. Firstly, the mesh generated for the rod is converted from the type T4 to the type T10 by inserting additional nodes at the midpoints of the edges with the utility T4_to_T10.

```
0042      [fens,gcells] = t4cylinderdel(t,R, nt,nR);
0043      [fens,gcells] = T4_to_T10(fens,gcells);
```

Secondly, the integration rule is boosted to a four-point formula.

```
0049      feb = feblock_defor_ss (struct ('mater',mater,
              'gcells',gcells,...
0050          'integration_rule',tet_rule (4),
              'Rm',@twist));
```

---

[4]Folder: SOFEA/examples/stress

Otherwise, the two functions are identical. The results are very different though. The higher order tetrahedron converges very quickly and produces monotonically converging answer for the lowest frequency (approximately 1766.6Hz). The convergence behavior with respect to the number of elements radially and longitudinally is compared with the earlier analysis with T4 in Fig. 13.8.

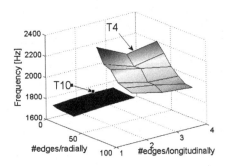

Fig. 13.8   Convergence of the lowest natural frequency. Comparison of the two tetrahedral elements.

13.5   Static analysis with hexahedra H8 and H20

13.5.1   *Hexahedron H8*

This element is a straightforward extension of the Q4 quadrilateral to three dimensions. In fact, the Q4 quadrilateral is compatible with the discretization on the faces of the hexahedron, which is essential for the implementation of the integrals of the surface terms (heat flux or traction loads).

The numbering of the nodes is given in Fig. 9.10. The basis function N_1 may be written as the product of one one-dimensional Lagrange interpolation function on the interval $-1 \leq \xi \leq +1$, one on the interval $-1 \leq \eta \leq +1$, and one on the interval $-1 \leq \zeta \leq +1$; all functions correspond to the left-hand side end of the interval

$$N_1(\xi, \eta, \zeta) = \frac{\xi - 1}{-1 - 1} \times \frac{\eta - 1}{-1 - 1} \times \frac{\zeta - 1}{-1 - 1} = \frac{(\xi - 1)(\eta - 1)(\zeta - 1)}{8} . \quad (13.2)$$

Analogously for the remaining functions: see the class method **bfun**[5] of the

---

[5]Folder: `SOFEA/classes/gcell/@gcell_H8`

class gcell_H8.

Gauss integration at $2 \times 2 \times 2$ points is considered a "full integration" of the stiffness (conductivity) matrix. There are also other, more economical integration rules (for instance, six-point mid-face rule). One-point integration at the centroid is not sufficient, and leads to rank-deficient element matrices.

13.5.2 Dilatational locking

Consider the split of the strains into two groups: relative change of volume, and distortional strain. The relative change in volume ϵ_v (*volumetric strain*, also called dilatational strain) may be written as

$$\epsilon_v = m^T \epsilon , \tag{13.3}$$

where we use the notation of Reference [Zienkiewicz, Taylor (1989)], namely the operator

$$[m]^T = \begin{bmatrix} 1\,1\,1\,0\,0\,0 \end{bmatrix} .$$

The part of the strain that corresponds to volume change is therefore

$$\epsilon_v = \frac{1}{3} mm^T \epsilon .$$

Correspondingly, the *deviatoric* (distortional) *strain* is

$$\epsilon_d = \epsilon - \epsilon_v = \epsilon - \frac{1}{3} mm^T \epsilon = I_d \epsilon , \tag{13.4}$$

where we use the deviatoric projector

$$I_d = 1 - \frac{1}{3} mm^T .$$

For *isotropic* elastic materials, applying the split of the strains in the constitutive equation, $\sigma = D\epsilon$, leads after some manipulations to the form that separates the two kinds of effects also in the stress

$$\sigma = D\epsilon = Km\epsilon_v + 2GI_0\epsilon_d = Kmm^T \epsilon + 2GI_0 I_d \epsilon , \tag{13.5}$$

where $K = E/3/(1 - 2\nu)$ is the *bulk modulus*, $G = E/2/(1 + \nu)$ is the shear modulus, and

$$I_0 = \text{diag} \left[1, 1, 1, \frac{1}{2}, \frac{1}{2}, \frac{1}{2} \right] .$$

As the Poisson's ratio approaches $\nu \to 1/2$, the elasticity model describes deformation which approaches the state of zero volume change. In other words, most (in the limit $\nu = 1/2$, all) of the deformation will be limited to distortion. Material models reproducing this behavior are called ***incompressible***. They are suitable for rubbery materials and fluid-saturated tissues, among others.

The bulk modulus approaches infinity for $\nu \to 1/2$. Therefore, the first part of the material stiffness matrix (13.5) grows without bounds as $\nu \to 1/2$. The consequence for the finite element formulation is that the large (and in the limit, infinite) stiffness $K\boldsymbol{m}\boldsymbol{m}^T$ will penalize any and all deformations that result in a non-zero relative volume change. Hence, unless the kinematics of deformation that a finite element provides allows for zero volumetric deformation to occur with $\boldsymbol{\epsilon} \neq \boldsymbol{0}$, the magnitude of possible strains will be severely limited, and in the worst case no deformation will be possible at all. Such a state of affairs is called ***dilatational locking***.

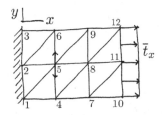

Fig. 13.9   Dilatational locking in a mesh composed of triangles.

Figure 13.9 supplies a visual explanation. Consider the case $K \to \infty$ with the mesh shown as a collection of triangles, but similar reasoning would apply to tetrahedra. The base in the triangle 152 is fixed. To conserve its volume, node 5 is allowed to move only vertically. Analogously for triangle 263, and to conserve the volume of triangle 256, the two nodes 5 and 6 must move by the same amount. Similarly we prove that node 9 must move only vertically. Compare these severe limitations on the deformation allowed by the incompressibility constrains with the expected shape produced by a tensile load in the x direction: clearly the results are not going to be very useful. In fact, it is easy to produce situations where the incompressibility would not allow any displacement at all at the nodes 2, 3, 5, 6, 9: For instance, consider the effect of applying a roller condition in the vertical direction on the upper edge, such as would be used to model an axis of symmetry.

The triangles T3 and tetrahedra T4 are in this way severely exposed to locking for almost incompressible materials. The deformations that they can experience are very limited as soon as we require zero (or very small) volumetric change.

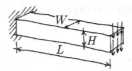

Fig. 13.10   Beam of almost incompressible material with transverse load.

Let us look at an example to study the performance of the hexahedron H8. Figure 13.10 should be referred to as to the geometry, but the important detail is that the material has a Poisson's ratio of $\nu = 0.4999$ (a number representative of some filled rubbers). (A comprehensive study is described later in Section 13.5.9; the analysis is run by the Matlab script rltb[6].) Figure 13.11 shows the coarsest and the finest mesh with a deflected shape.

Fig. 13.11   Meshes for the clamped beam.

The lower curve in Fig. 13.12 belongs to the regular H8, and it clearly illustrates that it also suffers severe dilatation locking (very slow convergence). However, in this case there is a remedy. The numerical integration of the stiffness matrix essentially imposes the constraint of (almost) zero volumetric strain at each integration point. Since this is impossible to meet for H8, the element locks. However, the technique known under the name of ***selective reduced integration*** [Hughes (2000)] may be applied to alle-

---

[6]Folder: SOFEA/examples/stress/3-D

viate the symptoms. Essentially, the two terms in (13.5) split the stiffness matrix, and the two parts of the stiffness matrix are integrated using different Gauss rules: the dilatational part is integrated with one-point rule, while a $2 \times 2 \times 2$ rule is used on the deviatoric part. Reducing the number of constraints due to lack of compressibility in this way leads to much improved performance. The upper curve in Fig. 13.12 is obtained with this formulation (H8 with selective reduced integration, SRI).

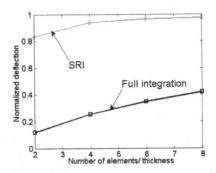

Fig. 13.12   Convergence of the tip deflection.

If we apply different numbers of elements along the axis of the beam, which results in elements elongated along the axis of the beam to varying degrees, we realize that while the problem of dilatation locking seems to have been solved, the bending response of the H8 element deteriorates significantly with the aspect ratio (the more elongated the element gets, the worse the result). There are evidently other effects than just dilatational locking at play here. Indeed, the H8 element suffers also from the so-called shear locking.

13.5.3   *Shear locking*

The representation of bending by the brick element H8 (or the quadrilateral Q4 in two-dimensional models) is notoriously poor due to excessive shear stiffness being generated as a side effect of the element assuming a "bent" configuration: refer to Fig. 13.13 for an illustration. The result is known as **shear locking**. It has to do with the partitioning of the deformation energy: consider a beam modeled with the brick element H8. When both the beam and the element are very stocky, the energies of shear and bending are comparable. On the other hand, when the beam and the element are very thin, the energy stored in bending should be much higher than that stored in shear. If the thin element needs to experience shear deformation to bend, the overall deformation will be severely limited. The reason is the big difference between the energy that should be stored in bending to attain a certain deflection (relatively small), and the spurious shear energy that is produced when the element is deformed to achieve that deflection (much larger). For a given amount of energy it then takes much smaller deflection to store it in shear than in bending, and locking results. We illustrate this behavior with an example next.

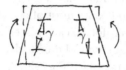

Fig. 13.13 Illustration of deformation that lead to shear locking.

13.5.4 *Thin clamped square plate with concentrated load*

The setup is illustrated in Fig. 13.14, and the input data is fully specified in the Matlab script `clsqconc`[7]. This is one of the classic benchmarks for thin structures, see the Reference [Timoshenko, Woinowsky-Krieger (1959)]. To match the thin-plate analytical solution is typically quite challenging for *plate* finite elements (that is structural elements specialized for bending deformations). Here we will attempt to compute this solution with *solid* elements. The essential boundary condition is trivial, but the concentrated force should give us a pause, as it is not an admissible load for 3-D elasticity. However, since the plate is thin, and since we are interested in the deflection of the structure as a "plate", committing this sin is not going to be fatal to the solution. Put differently, we are essentially pretending there is no singularity. Don't look at it: if you can't see it, it is not there.

Due to the symmetry of the problem, we will discretize just one quarter of the geometry, and apply symmetry boundary conditions. For a very thin plate (thickness/span = 1/2000), the deflected shape is shown in Fig. 13.15. However, this shape cannot be computed with the H8 element. Even with many thousands of elements, the normalized computed deflection is just a fraction of a percent of the analytical solution. The H8 element locks very

---

[7]Folder: `SOFEA/examples/stress/3-D`

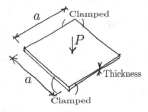

Fig. 13.14   Clamped square plate with center load.

badly: As the plate is so thin, most of the energy should be in bending, but the element deformation mode puts most of it into shear and as a result the model is way too stiff. Even for thicker plates (up to the thickness/span ratio of 0.1, when the plate should be considered thick), the element H8 delivers only middling accuracy: see Fig. 13.16. The error is unacceptably large not only for extremely thin plates, but also for plates which are quite common in structures (thickness/span = 1/100). In comparison, the tetrahedron T10 is a solid performer, even for plates which are rather thin. The mesh shown in Fig. 13.15 gives decent results for the shown thickness to span ratio.

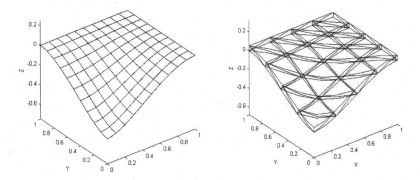

Fig. 13.15   Clamped square plate with center load. Very thin plate (thickness/span = 1/2000) and moderately thin (thick?) plate (thickness/span = 1/40); H20 and T10 meshes.

13.5.5   *Quadratic element H20*

Similarly to the pair T4 and T10, the quadratic hexahedron H20 cures most of the stiffness in the joints of the H8. Contrary to (likely) expectations,

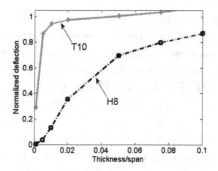

Fig. 13.16 Clamped square plate with center load. Normalized deflection under the load versus the thickness/span ratio, with a single element through the thickness and 20x20 elements in the plane. H8 and T10 meshes.

the quadratic element is not going to be derived by taking the Cartesian product of the basis functions of the quadratic one-dimensional element L3 (Section 9.2). The result would have been an element with 27 nodes: $3 \times 3 \times 3 = 27$, the so-called quadratic Lagrangean hexahedron.

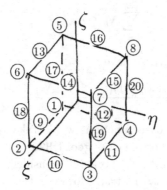

Fig. 13.17 Numbering of the nodes of the hexahedron H20.

The quadratic hexahedron H20 is a member of the **serendipity** family. The nodes associated with basis functions are located at the corners and the midpoints of the edges: see Fig. 13.17; eight corners plus the midpoints of 12 edges gives 20 basis functions. The derivation of the basis functions may proceed as follows: start with the basis functions of the hexahedron H8, and add in the quadratic functions associated with the midpoints. Then modify the original linear functions so that all the basis functions add up to one at any point within the element.

The functions that are being added for the midpoints are produced as products of one quadratic function (in one variable) and two linear functions (in the remaining two variables). For instance, consider the basis function N_{11}: the variation of this function along the edge $3, 4$ should be the quadratic $(1 - \xi^2)$ (Fig. 13.18). Furthermore, N_{11} should vanish at all the nodes except 11. The function $(1 - \xi^2)$ is zero along the faces $2, 3, 7, 6$ and $4, 1, 5, 8$. To make it vanish also along the two faces $7, 8, 5, 6$ and $1, 2, 6, 5$, we multiply it with the two functions $(\eta + 1)$ and $(\zeta - 1)$. Finally, this product is normalized to assume value $+1$ at the node 11. The result is

$$N_{11} = \frac{(1 - \xi^2)(\eta + 1)(\zeta - 1)}{8} \ .$$

The basis functions for all the other mid-side nodes are obtained in the same fashion. The final step is to subtract half of each mid-side basis function from the two corner basis functions that are borrowed from H8 along that edge so that we recover the partition of unity property (6.15)

$$\sum_{k=1}^{20} N_k(\xi, \eta, \zeta) = 1 \ .$$

Thus, for instance for N_1 we have

$$N_1 = \frac{(\xi - 1)(\eta - 1)(\zeta - 1)}{8} - \frac{N_{11}}{2} - \frac{N_{10}}{2} - \frac{N_{19}}{2} \ .$$

Detailing all the other basis functions would take up too much space: see the class method **bfun**[8] of the class **gcell_H20**.

Gauss integration at $3 \times 3 \times 3$ points is considered a "full integration" of the stiffness (conductivity) matrix. However, the theoretically insufficient Gauss scheme $2 \times 2 \times 2$ points often leads to much improved results, giving a less constrained model with considerably improved flexibility.

The clamped plate problem is revisited with the 20 node hexahedron: Figure 13.19 shows the results for the normalized deflection under the load for the very thin plate. The hexahedron H20 is used with the two Gauss quadratures. While the convergence is similar, the reduced $2 \times 2 \times 2$ scheme has an edge. The mesh shown on the left in Fig. 13.15 is good for approximately 98% accuracy with the reduced integration, and a few percent worse

---

[8]Folder: SOFEA/classes/gcell/@gcell_H20

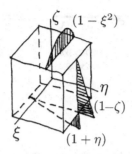

Fig. 13.18 Visualization of the three functions that give the basis function N_{11} for the hexahedron H20.

for the full integration. The quadratic tetrahedron T10 does not manage to produce acceptable results for plate this thin.

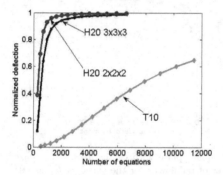

Fig. 13.19 Clamped square plate with center load. Normalized deflection under the load versus the number of unknowns for the thickness/span ratio= 1/2000, with a single element through the thickness and $4, 5, \ldots, 18$ elements along the in-plane directions.

The aspect ratios of the elements obviously improve for thicker plates. Figure 13.20 demonstrates that accuracy improves, and Fig. 13.21 shows that, given the change of the vertical scale, the results for the moderately thick plate are quite satisfactory for all elements. However, the tetrahedron will always be more expensive than the hexahedron. For instance, for a given level of accuracy, more degrees of freedom need to be included in the model; conversely, given a fixed budget of elements will lead to worse accuracy when using tetrahedra.

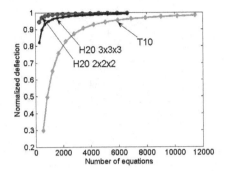

Fig. 13.20 Clamped square plate with center load. Normalized deflection under the load versus the number of unknowns for the thickness/span ratio= 1/200, with a single element through the thickness and $4, 5, \ldots, 18$ elements along the in-plane directions.

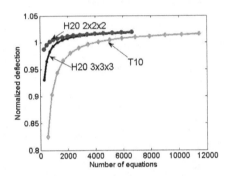

Fig. 13.21 Clamped square plate with center load. Normalized deflection under the load versus the number of unknowns for the thickness/span ratio= 1/40, with a single element through the thickness and $4, 5, \ldots, 18$ elements along the in-plane directions. The deflections are normalized by the analytical solution obtained for a thin plate, i. e. without consideration of shear deformations. That is why the numerical results apparently converge to a deflection higher than that predicted analytically.

13.5.6 *Quadratic element Q8*

The quadratic quadrilateral Q8 is compatible with the faces of the brick element H20 (compare Fig. 13.17 with Fig. 13.22). Therefore, to write down the basis functions for the quadrilateral, we may for instance substitute $\zeta = -1$ into the basis functions $N_j, j = 1, 2, 3, 4$ (corner functions), and $N_j, j = 9, 10, 11, 12$ (mid-side functions) of the hexahedron.

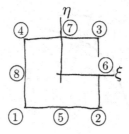

Fig. 13.22 Numbering of the nodes of the quadrilateral Q8.

13.5.7 *Pinched cylinder*

Now we'll have a look at some shell problems. The first structure is defined in Fig. 13.23, and the input data is available in the Matlab script `pinchcyl`[9]. This is a widely used benchmark for shell structures. Due to symmetry, only 1/8 of the full structure is discretized, with only a single element through the thickness (the resolution in the circumferential and longitudinal direction is much more important).

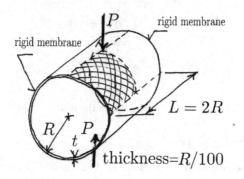

Fig. 13.23 Pinched cylinder: description of the problem.

Of the two quadratic elements, H20 and T10, only the hexahedron performs well, and only for the reduced integration scheme $2 \times 2 \times 2$; full integration produces a model which is too stiff. The results are summarized in Fig. 13.24.

The shape of the structure under the load is displayed in Fig. 13.25 and it is possible to discern the interplay between membrane and bending action. Also note that the elements are curved in their reference form.

---

[9]Folder: SOFEA/examples/stress/3-D

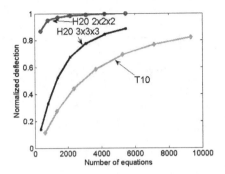

Fig. 13.24   Pinched cylinder. Normalized deflection under the load with elements H20 and T10.

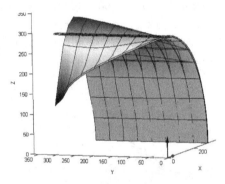

Fig. 13.25   Pinched cylinder. Shape under the load magnified 100,000 times.

13.5.8   *Pinched sphere*

This benchmark problem is defined in Fig. 13.26: Note that the structure (boundary value problem) as defined is free-floating, loaded with self-equilibrated forces. The input data is fully described in the Matlab script pinchsphere[10]. This benchmark is used to measure the capability of finite elements to model inextensional bending in shell structures. Due to symmetry, only 1/4 of the full structure is discretized with the hexahedron H20, with only a single element through the thickness. The geometry of the spherical shell is captured in two ways: (i) "flat", where only the corner nodes have been forced to lie on co-spherical surfaces; and (ii) "fully curved", where the element nodes have all been moved so that all nodes

---

[10]Folder: SOFEA/examples/stress/3-D

are located on co-spherical surfaces.

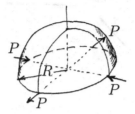

Fig. 13.26   Pinched sphere: description of the problem.

The deformed shape of the structure is shown for the two element configurations in Fig. 13.27. Clearly, the fully curved elements afford better accuracy, which is not entirely surprising: when the elements are assembled into a faceted surface, the structure is effectively stiffened ("corrugated"). However, modeling curved surfaces by faceting is routinely done, since in the limit the faceting becomes less important. Indeed, we may observe in Fig. 13.28 that both configurations lead to satisfactory convergence, and that the curving of the elements is important only for coarse meshes.

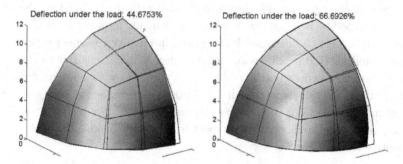

Fig. 13.27   Pinched sphere. Shape under the load magnified 10 times. Configuration of elements: flat (left), fully curved (right).

13.5.9   *Beam deflection revisited*

Finally, we return to the problem of the (almost) incompressible beam. We study a few models with the goal of understanding their properties concerning both dilatational and shear locking (including sensitivity to the height versus length ratio). (The input data is to be found in the Matlab script

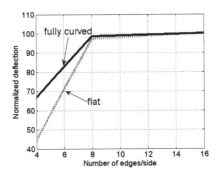

Fig. 13.28   Pinched sphere. Convergence of the deflection under the load for the two different configurations of the elements.

`rltb`[11].) The meshes have different aspect ratios, see Fig. 13.29, and correspondingly we may expect to be able to test the bending response: elements insensitive to aspect ratio are preferable to those that are sensitive. At the same time, we are able to ascertain robustness with respect to dilatation locking since the Poisson's ratio is close to $1/2$.

The results are summarized in Fig. 13.29 (note that the vertical scale is different in each graph). The hexahedron H20 with the reduced quadrature $2 \times 2 \times 2$ points is a uniformly accurate element, apparently performing well with respect to dilatation locking, and insensitive to aspect ratio. Full integration for H20 produces an element slightly more sensitive to element elongation, and generally stiffer. The hexahedron H8 with selective reduced integration (SRI) deals well with dilatational locking, but is very sensitive to element elongation: its bending response is poor. The tetrahedron T10 is apparently well-behaved with respect to dilatation locking, but it is quite sensitive to the aspect ratio.

13.6   Errors, validation, and verification

Modeling physical events (for instance, the deflection of an airplane wing when the aircraft is turning is a physical event) on the computer may be described as shown in Fig. 13.30. In the first step, a ***physical event*** is idealized into a ***mathematical model.*** The mathematical model has rarely exact (analytical) solutions, which gives rise to the need for a ***discrete model*** (the Galerkin finite element method in this book, but there are

---

[11]Folder: `SOFEA/examples/stress/3-D`

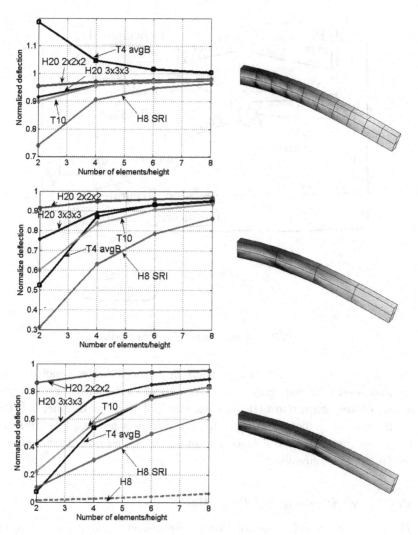

Fig. 13.29 Comparison of normalized deflections for the clamped beam with different aspect ratios. Top to bottom: 1:2, 1:5, 1:10 (height:length).

many others: finite difference and finite volume methods, boundary element methods, and so on). The unknowns in the discrete model are solved for using various numerical methods – solvers for systems of linear and nonlinear algebraic equations, eigenvalue solvers, integrators for systems of ordinary differential equations, ... – resulting in the ***solution*** of the dis-

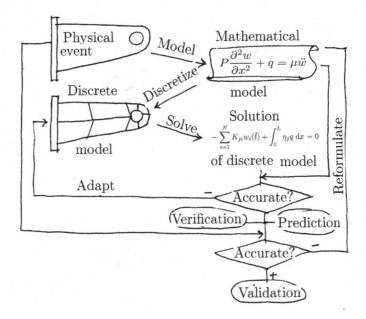

Fig. 13.30   Diagram of the modeling pipeline.

crete model. The accuracy of the solution to the discrete model may be then assessed with respect to the mathematical model with various methods of **error estimation** (Richardson extrapolation, for instance). If the accuracy is not sufficient, a better discrete model is produced by **adaptation**, else the solution is deemed acceptable for either one of two actions: prediction or verification.

13.6.1   *Verification and Prediction*

There are two uses for the solution of the discrete model, depending on whether a reference (exact, or very accurate) solution of the mathematical model is available or not. If it is available, we obtain a **verification** of the way the discrete model is implemented in the computer: the ingredients that go into the discrete model are correct, and the equations are solved right, therefore we observe convergence towards (closeness to) the reference solution of the mathematical model. The verification should be addressed thoroughly by the designers of the computational software, but users also tend to perform verifications to gain confidence in the software. Physical events are typically very idealized to produce mathematical models that can be solved very accurately or analytically. Such mathematical models are called **benchmarks**.

On the contrary, if the reference solution of the mathematical model is not known, the solution to the discrete model will make **predictions** of various quantities possible (deflections, natural frequencies, strains or stresses, energy, and so on).

13.6.2 *Validation*

Finally, if comparison of some quantities from the solution with **observations** of the corresponding quantities in the physical event is possible, and provided the agreement is good, we call the modeling pipeline **validated** for this particular physical event; otherwise, if an improvement to the mathematical model may be made, for instance by including aspects of the physical event that have been deliberately neglected before, we perform yet another adaptation and another pass through the modeling pipeline. If there is a significant mismatch, we may decide that a completely different mathematical model needs to be formulated, perhaps even requiring a new theory. In this way, we get the chance to **falsify** a theory.

13.6.3 *Errors*

The contributions to the detected mismatch are associated with the arrows in the graph of Fig. 13.30: there's the **modeling error** that accompanies the idealization of the physical event into a mathematical model, the **discretization error** when converting the mathematical model into a discrete model, the **solution error** produced by the numerical algorithms, and there may also be an **observation error** due to inaccurate or erroneous measurements. Finally, an important source of mismatch may also be **data uncertainty**, as all the input quantities will only be known with a certain margin of error (material parameters, geometric dimensions, and so on).

13.6.4 *Using modeling to make predictions*

Comparing the solution of the discrete model with experimental observations allows us to call the solution validated with respect to the specific physical event. If we needed to observe the physical event each time we

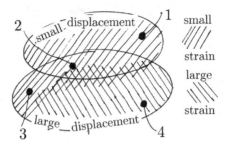

Fig. 13.31 Classes of stress analysis problems based on the magnitude of the displacement, and the magnitude of the strain. Examples: 1: concrete dam; 2: thin clamped plate under a transverse load; 3: steel measuring tape; 4: penetration of a steel slab by a high-velocity tungsten projectile.

ran a simulation so that it could be validated, there would not be many incentives for using the simulation in the first place. However, in practice we use validations for a series of specific events to sample the "event space" to establish a *range of validity*. If a particular physical event $\mathcal{E}$ resembles other events for which the validity of the model has been established, we assume that the model will be likely validated also for $\mathcal{E}$. As an example, consider the classification of the physical events with respect to the magnitude of displacements and strains (Fig. 13.31). If the physical event is the deformation of a structure that works roughly as a steel measuring tape, we assume that it is in the class of events for which models based on large displacements but small strains are validated. Note that the two model classes, small and large displacement, overlap. That is because there's no sharp division of the physical events that can be modeled with either depending on the required accuracy and the importance of capturing the effect of large displacements. For instance, deflections of clamped plates under transverse loads depend to a certain degree on the tension that develops in the structure when it deforms; if the contribution is negligible, small-displacement model could be adequate, otherwise a large-deflection model may be required.

13.6.5 *Using benchmarks*

The value of benchmarks for the verification of numerical models is clear, but there is another reason analysts should take advantage of benchmark problems: they need to build up a feel for the relative accuracy and robustness of finite elements, and such data is readily accessible in convergence studies performed on benchmark problems because of the availability of the reference solution. The convergence graphs displayed in the previous sections of this chapter are a source of valuable insights.

Eventually, analysts develop intuition to help them assess the suitability of a particular discretization for a particular physical event. For example, performing a series of convergence analyses for different values of the Poisson ratio will help us create a map of the performance of a particular element as it depends on this parameter. Figure 13.32 shows such a map for the fully-integrated brick element H8.

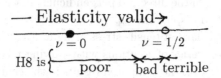

Fig. 13.32   The effect of the Poisson ratio on the accuracy of the fully-integrated brick element H8.

Figure 13.33 illustrates on the example of plate-like structures how their thickness would affect the choice of the mathematical model as well as the choice of the discretization. We see how for relatively thicker structures the solid elements might be acceptable discretization, while for really thin plates a specialized plate element (based either on the shear-free Kirchhoff theory, or based on the Mindlin theory that incorporates transverse shear) would be more effective.

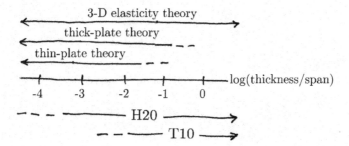

Fig. 13.33   Schematic classification of stress and deflection analysis finite elements for plate-like structures.

Exercises

(1) Chart 4.34 from Peterson's Stress Concentration Factors [Pilkey (1997)] shows the stress intensity factors for an infinite thin slab (membrane) with an infinite row of circular holes loaded with biaxial tensile distributed forces. Consider $d = 10$mm, and thickness of the slab 2mm; assume that the slab is of AISI 1005 Steel.

   (a) Use as many planes of symmetry as possible and the St. Venant's principle to reduce the domain to finite size. Justify your choices, and illustrate the boundary conditions with a sketch. Hint: Apply the tractions in the direction perpendicular to the symmetry planes by prescribed displacements.

   (b) Compute the largest tensile stress using Richardson's extrapolation applied to a series of solutions with varying mesh sizes for $d/\ell = 0.1, 0.2, 0.4$. Plot the computed results in the chart 4.34 to compare with the analytical solutions.

   (c) Reduce the size of the domain so that it fits into a cube of side $\ell/2$, and recompute the largest tensile stress for $d/\ell = 0.4$. Evaluate the effect of the size of the domain on the computed tensile stress.

Chapter 14

Analyzing the Stresses

Stresses are often of primary interest when designing structures. However, there are some inherent limitations to what is computable and how.

For example, when analyzed with the elasticity model, some features generate stresses that are infinite at some points. Obviously, there is little merit in attempts aimed at computing the stress at such points. Even away from these points, without special models the stress is often very inaccurate. Therefore, it behooves us to understand which features lead to infinite stresses so that we can formulate solution strategies appropriately.

14.1 Singularities

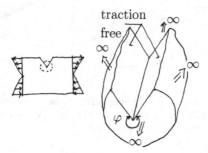

Fig. 14.1   Notch and the idealized geometry of an infinite wedge.

A set of geometric features leads to states of stress which can be analyzed with two dimensional models (the displacements are functions of two variables). The idealization of such geometries is the *wedge* as a locus of points where two bounding surfaces meet at an angle. Figure 14.1 shows a geometry with a notch (even though displayed in cross-section, the notch needs to be understood as an edge). Away from the locations where the notch edge runs out into the side surfaces, the state of stress may be analyzed using the idealization shown on the right, the *wedge* [Barber (1999)]. The material surrounding the edge extends to infinity in all directions, but the solution is of interest only in the immediate vicinity of the edge. The analytical solution for symmetric and anti-symmetric loadings possesses for angles $\varphi > 180°$ a singularity in the stress components of the form

$$\sigma \sim r^\alpha , \tag{14.1}$$

where r is the radial distance from the edge, and $\alpha \leq 0$ measures the strength of the singularity. Figure 14.2 provides a sketch of the dependence of the exponent α on the angle φ. For symmetric loads, even a very slight notch will lead to singular stresses, and the strongest singularity occurs for $\varphi = 360°$ (infinitely sharp crack). Similarly, anti-symmetric loading will produce the same strength of singularity for a crack configuration. In the same figure on the right, the graph shows that the stronger the singularity, the wider the general area around the wedge front where the stresses are high (they are infinite at the edge $r = 0$ for all strengths).

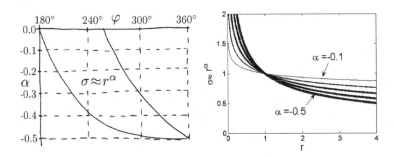

Fig. 14.2   On the left: Strength of the singularity in the stress. Symmetric (lower curve) and anti-symmetric (upper curve) loading. On the right: Variation in the radial direction in dependence on the strength of the singularity.

There are also other sources of singularities in the elasticity model. Singular stresses accompany edges along ***multi-material interfaces***, as shown in Fig. 14.3, or in Fig. 14.5. Singular stresses are also generated at sudden transitions from prescribed displacement to prescribed traction (***discontinuity in boundary conditions***), see Fig. 14.4.

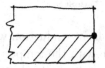

Fig. 14.3   Singularity at a multi-material interface.

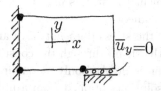

Fig. 14.4   Singularities due to sudden change in displacement boundary conditions.

We also pointed out in Section 11.6.4 that **concentrated loads** at points or along curves generate infinite stresses. As discussed in detail there, the energy in the model is infinite in the limit. Therefore, this kind of singularity is "even worse" than the singular stresses near the tip of a wedge. Also, point supports and supports along curves leads to infinite stresses with similar characteristics as concentrated loads (unless the associated reactions are identically zero).

The most complex singularities to analyze are the true **3-D singularities at corners**. Examples include (call for Fig. 14.5) the intersections of wedge fronts (crack fronts) with free surfaces (marked 1), corners of multi-material wedges (2), corners on multi-material interfaces (3), or corners in homogeneous geometries (4,6), conical points (marked 5), or corners in wedges (on crack fronts) (marked 7).

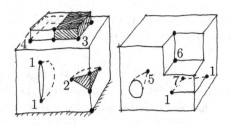

Fig. 14.5   3-D singular points.

14.2   Interpretation of stresses

Similarly to the problem of heat conduction where the heat flux is defined only through its relationship with the temperature gradient, the stress distribution is detectable in elasticity finite element models by recourse to the strains. The strains are defined as the spatial derivatives of the displacements, and hence it makes sense to talk of some degree of continuity. For instance, across an element the strains may be constant (T4 tetrahedra), or vary linearly (T10 tetrahedra).

On the other hand, the stress does *not* have any such continuity. As the stress is related to the strain through the constitutive equation, and as the constitutive equation is invoked only at quadrature points, the stress is consistent with the solution obtained from a finite element model *only* at the quadrature points. This is especially true when a more complex relationship exists between the strains and the stresses, for instance in plasticity or viscoelasticity. In the present model of elasticity, it may seem attractive to use the constitutive equation at any point within the element to compute the stresses. However, it needs to be realized that such stresses are an *interpretation* of the solution, not the solution itself.

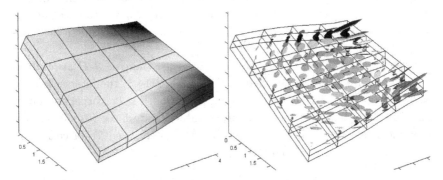

Fig. 14.6   Interpretations of the calculated stresses. Stress τ_{yz}. Left: extrapolated, continuous field; Right: stress ellipsoids at integration points.

As an illustration, consider Fig. 14.6: on the right, the stresses consistent with a finite elements solution are displayed at the $2 \times 2 \times 2$ quadrature points as ellipsoids that are visual representations of the principal stresses and directions of the principal stresses. On the left, an interpretation of the distribution of stresses is produced as a continuous map, where the stress magnitude is calculated at each node and then interpolated using the element basis functions. The magnitude of the stress at a node is computed by extrapolating from the magnitude of the stress at the quadrature points nearby; in this case, by taking an inverse-distance-weighted average of the stresses at all the quadrature points in all the elements connected to the node. The usefulness of such a visual representation of the stress distribution for the evaluation of the results is evident, but the fact that such a continuous stress field is an interpretation of the finite element results (often performed in a so-called *postprocessing* step) must be taken into account. In particular, the stresses are typically *smoothed* in these postprocessing procedures and spikes in stress may get wiped out. It should also be realized that *extrapolating* the stress may be fraught with significant errors due to overshoot or undershoot.

14.3 Stress concentrations

The finite element formulations in this book may be applied to the solution of the so-called stress concentration problems. These are produced by stress raisers, which are features that cause local or global increases in the stress, but which do not lead to infinite stresses. The book [Pilkey (1997)] is an extremely useful resource; a mode of operation where the solution of a finite element model is cross-checked with this book, or vice versa, is to be generally recommended.

To be aware of stress raisers is highly advisable for all modelers, since stress concentrations where the stress changes rapidly are generally locations of high error in the finite element solution: compare with Section 10.1.3. Therefore, appropriately constructed mesh will reflect the presence of stress raisers by suitable refinement (reduction in mesh size) around the stress concentrations. Figure 14.7 presents examples of stress raisers.

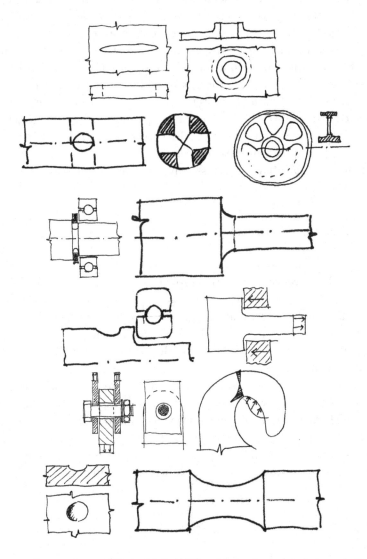

Fig. 14.7   Stress raisers. Elliptical hole, stiffened hole, through-hole in a shaft, mass-reduction holes in a flywheel, snap-on ring groove, shoulder fillet, stress relief groove, T-head suspension, clevis and lug joint, curved hook, depression in a plate, cross-section reduction in a shaft.

Chapter 15

Plane Strain, Plane Stress, and Axisymmetric Models

In this chapter we will reduce the three-dimensional elastodynamic model to require solutions in terms of only two space variables (and the time). This is possible by introducing various assumptions for strains or stresses (and some restrictions on the form of the constitutive equation and loads).

15.1 Plane strain model reduction

The starting point is the observation that for some solids of uniform cross-section (right angle cylinders), such as the one shown in Fig. 15.1, the set of *assumptions* that **(i)** $u_z(x, y, z) = 0$, and $u_x = u_x(x, y)$, $u_y = u_y(x, y)$ seems to approximate the deformation well. As examples consider a long pipe under internal pressure (well removed from any valves or caps), or a straight concrete dam, or the midsection of a wall panel. The reduction to two space variables will be possible if the third balance equation is satisfied

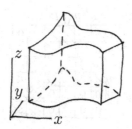

Fig. 15.1   Right-angle cylinder.

by design

$$\frac{\partial \tau_{zx}}{\partial x} + \frac{\partial \tau_{zy}}{\partial y} + \frac{\partial \sigma_z}{\partial z} + \bar{b}_z = \rho \ddot{u}_z \,.$$

The identically vanishing u_z also implies $\epsilon_z = 0$. Furthermore, we have for the shear strains

$$\gamma_{zx} = 0, \quad \gamma_{zy} = 0 \,.$$

Provided the constitutive equation allows, the conjugate shear stresses may also vanish. The condition for τ_{zx} is **(ii)**

$$\tau_{zx} = [\boldsymbol{D}]_{5,(1:6)} [\boldsymbol{\epsilon}] = 0 \,,$$

where $[\boldsymbol{D}]_{5,(1:6)}$ is taken to mean columns 1 through 6 of row 5. This condition is satisfied provided the coefficients of the material stiffness that multiply the nonzero strains are zero,

$$D_{51} = D_{52} = D_{54} = 0 \,.$$

Symmetry also implies $D_{15} = D_{25} = D_{45} = 0$. Analogously, for the stress τ_{zy} the conditions on the material stiffness coefficients applies to row 6 (column 6). The situation is illustrated in Fig. 15.2: coefficients which could be nonzero are hatched, coefficients which must be zero are marked as such, and coefficients which are probably best set to zero are blanks (those would couple τ_{xy} and σ_z).

The general *orthotropic* material of Section 12.5.2 complies with these conditions provided **(iii)** one of the orthotropy axes is aligned with the $z-$axis. A more general material model could be devised, but for practical purposes it is sufficient to consider the orthotropic material.

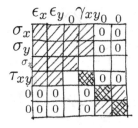

Fig. 15.2   Entries of the material stiffness matrix for plane strain.

As a consequence of the assumptions on the displacements, and of the constraints on the form of the material stiffness, the third equilibrium equation becomes

$$\frac{\partial \sigma_z}{\partial z} + \bar{b}_z = 0 \ .$$

But since we already have $\epsilon_x = \epsilon_x(x, y)$ and $\epsilon_y = \epsilon_y(x, y)$, we can write

$$\sigma_z = [\boldsymbol{D}]_{3,(1:2)} \begin{bmatrix} \epsilon_x \\ \epsilon_y \end{bmatrix} \ ,$$

and we see that $\sigma_z = \sigma_z(x, y)$ provided **(iv)** all the coefficients $[\boldsymbol{D}]_{3,(1:2)}$ and $[\boldsymbol{D}]_{(1:2),3}$ are functions of x, y but not of z. Then, the equilibrium equation simply becomes **(v)** a definition of the admissible loads

$$\bar{b}_z = 0 \ .$$

The boundary conditions consistent with the modeling assumptions and constraints **(i)-(v)** are

- top and bottom plane: $\bar{u}_z = 0$, $\bar{t}_x = 0$, $\bar{t}_y = 0$;
- cylindrical surface: mixture of essential and natural boundary conditions, where none of the prescribed components depend on z.

The initial conditions satisfy the modeling assumptions and constraints **(i)-(v)** provided the initial displacements and velocities do not depend on z.

Since $\epsilon_z = 0$, the constitutive equation is written for the nonzero normal stresses and the shear stress as

$$\begin{bmatrix} \sigma_x \\ \sigma_y \\ \tau_{xy} \end{bmatrix} = \begin{bmatrix} [\boldsymbol{D}]_{(1:2),(1:2)} & 0 \\ 0 & D_{44} \end{bmatrix} \begin{bmatrix} \epsilon_x \\ \epsilon_y \\ \gamma_{xy} \end{bmatrix} \ . \tag{15.1}$$

The stress-divergence operator for the plane strain model with just two balance equations and three stresses assumes the form

$$\mathcal{B}^T = \begin{bmatrix} \partial/\partial x & 0 & \partial/\partial y \\ 0 & \partial/\partial y & \partial/\partial x \end{bmatrix} \ . \tag{15.2}$$

The model for the fully three-dimensional elasticity resulted in the ODE system (12.31). In the present case, at this point we are still dealing with the same three-dimensional problem. However, when we substitute all the assumptions and constraints into the Eqs. (12.22), (12.24), (12.25), (12.26),

(12.28), and (12.30), the following observations are made. Firstly, the values of all the degrees of freedom in the direction of the z-axis are known to be zero. Secondly, all the loads in this direction are also zero. Consequently, the third equilibrium equation may be ignored, and the test and trial function will have only two components, x and y. Thirdly, all the volume integrals may be precomputed in the z direction as all the integrands are independent of z. Thus, for instance the stiffness matrix integral (12.29) may be written as

$$K_{(j,i)(k,m)} = \left[\int_V \mathcal{B}^T \left(N_j(\boldsymbol{x}) \right) \boldsymbol{D} \mathcal{B} \left(N_k(\boldsymbol{x}) \right) \, dV \right]_{im} = \qquad (15.3)$$

$$\left[\int_S \mathcal{B}^T \left(N_j(\boldsymbol{x}) \right) \boldsymbol{D} \mathcal{B} \left(N_k(\boldsymbol{x}) \right) \Delta z \, dS \right]_{im} . \qquad (15.4)$$

Here Δz is the thickness of the slab in the z direction, S is the cross-sectional area of the cylinder. Only σ_x, σ_y, τ_{xy} (and correspondingly ϵ_x, ϵ_y, γ_{xy}) matter, and the strain-displacement matrix is the transpose of (15.2).

15.2   Plane stress model reduction

The plane stress model idealizes the following situation: imagine a thin slab of uniform thickness (the technical term would be **membrane**) with both plane surfaces at the top and bottom (see Fig. 15.3) traction free, and the boundary conditions on the cylindrical surface independent of the z coordinate. Based on energy considerations we may presume that only the in-plane stresses would play a major role.

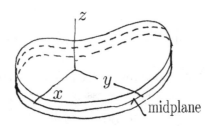

Fig. 15.3   Right-angle cylinder/thin slab/membrane.

Therefore, we formulate a model starting from **(i)** the assumptions $u_x = u_x(x,y)$, $u_y = u_y(x,y)$, and u_z symmetric with a respect to the midplane of the membrane. To reduce the problem to two dimensions, we have to try to get rid of the third equilibrium equation.

$$\frac{\partial \tau_{zx}}{\partial x} + \frac{\partial \tau_{zy}}{\partial y} + \frac{\partial \sigma_z}{\partial z} + \bar{b}_z = \rho \ddot{u}_z \ .$$

Neither of the stresses in it will be exactly zero, since the corresponding strains are in general nonzero. Nevertheless, making an inference from the uniform, and small, thickness of the membrane, we adopt yet another assumption, (ii) $|\tau_{zx}| \ll 1$, $|\tau_{zy}| \ll 1$, and $|\sigma_z| \ll 1$. Next, (iii) the through-the-thickness body load component is assumed to be zero, $\bar{b}_z = 0$. Finally, in order for the third equilibrium equation to be truly negligible, (iv) we invoke the symmetry with respect to the midplane, and establish that the integral of this equation through the thickness of the membrane will vanish, i.e. the resultant perpendicular to the plane of the membrane will vanish

$$\int_{-h/2}^{h/2} \frac{\partial \tau_{zx}}{\partial x} + \frac{\partial \tau_{zy}}{\partial y} + \frac{\partial \sigma_z}{\partial z} + \bar{b}_z \ \mathrm{d}z = \int_{-h/2}^{h/2} \rho \ddot{u}_z \ \mathrm{d}z = 0 \ .$$

Furthermore, in order to comply with the symmetry requirement, we shall assume that (v) the material is orthotropic, with one orthotropy axis perpendicular to the midplane. The material stiffness matrix will therefore have the appearance shown in Fig. 15.4.

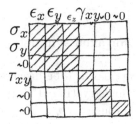

Fig. 15.4   Entries of the material stiffness matrix for plane stress.

As the last step, the strain ϵ_z is eliminated from the constitutive equation. Assuming $\sigma_z \approx 0$, we have

$$\sigma_z = [D]_{3,(1:3)} \begin{bmatrix} \epsilon_x \\ \epsilon_y \\ \epsilon_z \end{bmatrix} \approx 0 \ ,$$

which gives

$$\epsilon_z = -D_{33}^{-1}[\boldsymbol{D}]_{3,(1:2)} \begin{bmatrix} \epsilon_x \\ \epsilon_y \end{bmatrix}.$$

Therefore, we obtain

$$\begin{bmatrix} \sigma_x \\ \sigma_y \end{bmatrix} = \left[[\boldsymbol{D}]_{(1:2),(1:2)}, [\boldsymbol{D}]_{(1:2),3} \right] \begin{bmatrix} \epsilon_x \\ \epsilon_y \\ \epsilon_z \end{bmatrix} =$$

$$\left([\boldsymbol{D}]_{(1:2),(1:2)}, -D_{33}^{-1}[\boldsymbol{D}]_{(1:2),3}[\boldsymbol{D}]_{3,(1:2)} \right) \begin{bmatrix} \epsilon_x \\ \epsilon_y \end{bmatrix}, \qquad (15.5)$$

which together with $\tau_{xy} = D_{44}\gamma_{xy}$ may be substituted into the first two equilibrium equations, rendering them functions of x, y only, provided **(vi)** the body load is $\bar{b}_x = \bar{b}_x(x, y)$, $\bar{b}_y = \bar{b}_y(x, y)$.

The boundary conditions consistent with the modeling assumptions and constraints **(i)-(vi)** are

- top and bottom plane traction-free ($\bar{t}_z = 0$, $\bar{t}_x = 0$, $\bar{t}_y = 0$);
- cylindrical surface: mixture of essential and natural boundary conditions, where none of the prescribed components depend on z.

The initial conditions satisfy the modeling assumptions and constraints **(i)-(vi)** provided the initial displacements and velocities do not depend on the z coordinate.

15.3 Model reduction for axial symmetry

The last model reduction approach to be discussed in this chapter, is the case of *torsionless* axial symmetry (Fig. 15.5). The main assumption is that **(i)** all planes passing through the y axis are symmetry planes. (The y axis will be referred to as the axis of symmetry.) Therefore, points in any particular cross-section move only in the radial and axial direction, and do not leave the plane of the cross-section (the circumferential displacement is identically zero). Furthermore, points on circles in planes perpendicular to the axis of symmetry, with centers on the axis of symmetry, experience the same radial and axial displacements.

Let us consider one particular cross-section, for instance as indicated in Fig. 15.5. Since the displacements in the plane of the cross-section are symmetric with respect to the axis of symmetry, we will consider only

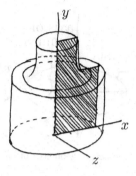

Fig. 15.5   Axially symmetric geometry.

the part to one side of the axis of symmetry (hatched in Fig. 15.5). The reduction from three equations of equilibrium to two equations in the plane of the cross-section, where x is the radial direction, and y is the axial direction, will succeed provided the equilibrium equation in the direction perpendicular to the plane of the cross-section (z) is satisfied. Thus, we consider

$$\frac{\partial \tau_{zx}}{\partial x} + \frac{\partial \tau_{zy}}{\partial y} + \frac{\partial \sigma_z}{\partial z} + \bar{b}_z = \rho \ddot{u}_z \, ,$$

where as the first simplification we note $u_z = 0 \rightarrow \ddot{u}_z = 0$. Furthermore, by symmetry we must have $\bar{b}_z = 0$.

Now for the stresses: As indicated in Fig. 15.6, by assumption the circles in planes perpendicular to the axis of symmetry are transformed by the deformation again into circles in planes perpendicular to the axis of symmetry. Therefore, right angles between the plane of the cross-section and the tangent to the circle where it intersects the cross-section will remain right angles, and the shear strains are $\gamma_{zx} = 0$, and $\gamma_{zy} = 0$. If the material is **(ii)** orthotropic, with one axis of orthotropy perpendicular to the cross-section (for instance a wound composite, with the reinforcing fibers running approximately in circles around the axis of symmetry), we may conclude that $\tau_{zx} = 0$, and $\tau_{zy} = 0$. Finally, since the material on a given circle experiences the same stress in all cross sections, $\partial \sigma_z / \partial z = 0$ (while in general $\sigma_z \neq 0$). Conclusion: the equation of motion in the z direction for each cross-section plane (and in particular the one in which the two-dimensional model is formulated) is satisfied exactly.

It remains to express the circumferential strain in terms of the displacements in the plane of the cross-section. By inspection of Fig. 15.6,

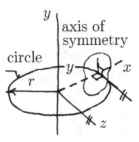

Fig. 15.6   Axially symmetric geometry: geometry of circles in planes perpendicular to the axis of symmetry.

displacement of the circle in the y direction does not change its circumference, while displacement radially means that the circle of radius x will experience strain

$$\epsilon_z = \frac{2\pi(x + u_x) - 2\pi x}{2\pi x} = \frac{u_x}{x}.$$

The strain-displacement operator for this model may therefore be written as

$$\mathcal{B} = \begin{bmatrix} \partial/\partial x & 0 \\ 0 & \partial/\partial y \\ 1/x & 0 \\ \partial/\partial y & \partial/\partial x \end{bmatrix}. \tag{15.6}$$

The stress divergence operator (the transpose of (15.6)) gives the reduced form for the equations of equilibrium (11.25)

$$\frac{\partial \sigma_x}{\partial x} + \frac{\partial \tau_{xy}}{\partial y} + \frac{\sigma_z}{x} + \bar{b}_x = \rho \ddot{u}_x$$

$$\frac{\partial \tau_{xy}}{\partial x} + \frac{\partial \sigma_y}{\partial y} + \bar{b}_y = \rho \ddot{u}_y$$

where the presence of σ_z should be noted: recall that the equation of motion in the radial direction holds for a wedge-shaped element, hence the contribution of the circumferential stress to the radial direction must be included.

     The constitutive equation is obtained from the full three-dimensional relationship by extracting appropriate rows and columns

$$
\begin{bmatrix} \sigma_x \\ \sigma_y \\ \sigma_z \\ \tau_{xy} \end{bmatrix} = \begin{bmatrix} [\boldsymbol{D}]_{(1:3),(1:3)} & 0 \\ 0 & D_{44} \end{bmatrix} \begin{bmatrix} \epsilon_x \\ \epsilon_y \\ \epsilon_z \\ \gamma_{xy} \end{bmatrix}. \tag{15.7}
$$

15.4   Material stiffness for two-dimensional models

The material stiffness matrices are (15.1) for plane strain, equation (15.5) for plane E stress, and (15.7) for axial symmetry. While all are for models reduced to two displacement components, the stiffness matrices are clearly different. The material class `mater_defor_ss_linel_biax` provides the tangent moduli by transforming the material stiffness of a three-dimensional material to the form suitable for a particular reduced two-dimensional model. The class method `tangent_moduli` first retrieves the three-dimensional material stiffness from the property object (line 0014), and then the switch statement implements the computation (or subsampling) expressed in the three formulas.

```
0013 function D = tangent_moduli¹(self, context)
0014     D= get(self.property,'D', context);
0015     switch self.reduction
0016         case 'axisymm'
0017             D =D(1:4, 1:4);
0018         case 'strain'
0019             D =D([1, 2, 4],[1, 2, 4]);
0020         case 'stress'
0021             Dt =D(1:2, 1:2)-D(1:2,3)*D(3,1:2)/D(3,3);
0022             D =D([1, 2, 4],[1, 2, 4]);
0023             D(1:2, 1:2)= Dt;
0024         otherwise
0025             error([' Reduction '
                          self.reduction ' not recognized']);
0026     end
0027 end
```

Note well that there is no error checking concerning the assumptions in the three models. In particular, all three assume there is no coupling between

---

[1]Folder: `SOFEA/classes/mater/@mater_defor_ss_linel_biax`

normal stresses and the in-plane shear stress. If the three-dimensional material stiffness was retrieved from a property class which did not comply with these requirements, incorrect results would not be unexpected.

15.5 Strain-displacement matrices for two-dimensional models

The three models need different strain displacement matrices. The plane strain/ planes stress defines the stress-displacement matrix as the transpose of (15.2)

$$\mathcal{B} = \begin{bmatrix} \partial/\partial x & 0 \\ 0 & \partial/\partial y \\ \partial/\partial y & \partial/\partial x \end{bmatrix} . \tag{15.8}$$

while the axially symmetric model works with (15.6). Consider the discretization of a particular structure with the triangles T3. In general, it is not possible to tell whether the model is axisymmetric, plain strain, or plain stress just by looking at the mesh. The finite elements (geometric cells) need to be told which they are. For instance, an axisymmetric triangle is created by the constructor when a flag is supplied:

```
gcell_T3(struct('conn', [2, 4, 7],'axisymm', true));
```

Based on this flag, the two-manifold class `gcell_2_manifold` decides which type of the strain-displacement matrix should be produced by the method `Blmat`.

```
0024 function B = Blmat²(self, pc, x, Rm)
0025     Nder = bfundpar (self, pc);
0026     Ndersp=bfundsp(self,Nder,x*Rm);
0027     nfens = size(Ndersp, 1);
0028     dim=length (Rm(:,1));
0029     if self.axisymm
0030         N = bfun(self, pc);
0031         xyz =N'*x;
0032         r=xyz(1);
0033         B = zeros(4,nfens*dim);
0034         for i= 1:nfens
```

---

[2]Folder: SOFEA/classes/gcell/@gcell_2_manifold

```
0035                    B(:,dim*(i-1)+1:dim*i)=...
0036                      [Ndersp(i,1) 0; ...
0037                       0            Ndersp(i,2); ...
0038                       N(i)/r 0; ...
0039                       Ndersp(i,2) Ndersp(i,1) ]*Rm(:,1:2)';
0040            end
0041        else
0042            B = zeros(3,nfens*dim);
0043            for i= 1:nfens
0044                    B(:,dim*(i-1)+1:dim*i)=...
0045                      [Ndersp(i,1) 0; ...
0046                       0            Ndersp(i,2); ...
0047                       Ndersp(i,2) Ndersp(i,1) ]*Rm(:,1:2)';
0048            end
0049        end
0050        return;
0051 end
0052
```

15.6 Integration for two-dimensional models

As discussed earlier, the volume integrals for plane strain, such as the one required for the stiffness matrix, incorporate the third dimension as a "thickness". Integration through the third dimension reduces to multiplication with the thickness.

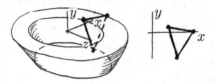

Fig. 15.7   Volume integration for axisymmetric analysis.

The volume integrals for the axial-symmetry model reduction approach differ from plane stress or plane strain in that the "for free" integration is performed along the circumference of the rotationally symmetric body, not "through the thickness". For instance, the stiffness matrix integral (12.29) may be written as

$$K_{(j,i)(k,m)} = \left[\int_V \mathcal{B}^T \left(N_j(\boldsymbol{x}) \right) \boldsymbol{D}\mathcal{B} \left(N_k(\boldsymbol{x}) \right) \, \mathrm{d}V \right]_{im} =$$
$$\left[\int_S \mathcal{B}^T \left(N_j(\boldsymbol{x}) \right) \boldsymbol{D}\mathcal{B} \left(N_k(\boldsymbol{x}) \right) 2\pi x \, \mathrm{d}S \right]_{im}, \qquad (15.9)$$

where $2\pi x \, \mathrm{d}S$ is the volume element of the ring generated by revolving the area $\mathrm{d}S$ around the axis of symmetry y—compare with Fig. 15.7. The volume integration may be performed by numerical quadrature over the area S but using a *volume* Jacobian

$$\int_V f \, \mathrm{d}V = \int_S f 2\pi x \, \mathrm{d}S \approx \sum_{k=1}^{M} f(\xi_k, \eta_k) J_{\mathrm{vol}}(\xi_k, \eta_k) W_k ,$$

where the volume Jacobian is defined as ($\det [J(\xi_k, \eta_k)]$ is the surface Jacobian)

$$J_{\mathrm{vol}}(\xi_k, \eta_k) = 2\pi x(\xi_k, \eta_k) \det [J(\xi_k, \eta_k)] . \qquad (15.10)$$

In this way, all volume integrals are performed in exactly the same way, be they defined over a three dimensional manifold (solid), a two-dimensional manifold (surface equipped with a thickness, or swept around an axis of symmetry), a one dimensional manifold (curve equipped with a cross-sectional area), or a zero-dimensional manifold (point endowed with a chunk of volume). For the two-dimensional models of this chapter, the volume Jacobian is computed by the method `Jacobian_volume` defined for the class `gcell_2_manifold`. The Matlab code closely mirrors the formulas; the method `other_dimension` evaluates the thickness at the given integration point `pc` (which in the plane-stress case could be variable).

```
0015 function detJ = Jacobian_volume³(self, pc, x)
0016     if self.axisymm
0017         N = bfun(self,pc);
0018         xyz =N'*x;
0019         detJ = Jacobian_surface(self, pc, x)*2*pi*xyz(1);
0020     else
0021         detJ=Jacobian_surface(self,pc,x)
                  *other_dimension(self,pc,x);
0022     end
0023 end
```

---

[3]Folder: `SOFEA/classes/gcell/@gcell_2_manifold`

Traction loads, such as prescribed pressure on part of the bounding surface, needs to be evaluated with surface integrals. Quite analogously to the volume integration, the surface integrals are numerically evaluated along curves, but the Jacobians are surface Jacobians. Compare with Figs. 15.3 and 15.8: for instance, for axial symmetry the surface Jacobian is

$$J_{\text{surf}}(\xi_k) = 2\pi x(\xi_k) \det [J(\xi_k)] \ . \tag{15.11}$$

Here $\det [J(\xi_k)]$ is the curve Jacobian (6.58). For the two-dimensional models, the surface Jacobian is computed by the method `Jacobian_surface` defined for the class `gcell_1_manifold`.

```
0015 function detJ = Jacobian_surface⁴(self, pc, x)
0016     if self.axisymm
0017         N = bfun(self,pc);
0018         xyz =N'*x;
0019         detJ = Jacobian_curve(self, pc, x)*2*pi*xyz(1);
0020     else
0021         detJ = Jacobian_curve(self,pc,x)
                        *other_dimension(self,pc,x);
0022     end
0023 end
```

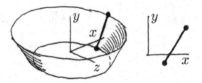

Fig. 15.8 Surface integration for axisymmetric analysis.

15.7 Thermal strains in two-dimensional models

The thermal strains must be of the form (12.33) as the two-dimensional models all require properly aligned orthotropic constitutive relations. In order to define the thermal strain loads (12.39), three ingredients are needed:

---

[4]Folder: `SOFEA/classes/gcell/@gcell_1_manifold`

the strain-displacement matrix, the material stiffness, and the thermal strains. The former two have been discussed before, it remains to outline the calculation of the thermal strains when the model is reduced.

The plane stress model and the axially symmetric model may refer directly to (12.33), because the stress in the former is independent of the strain in the third direction, while all strains are incorporated in the latter. On the other hand, in the plane strain model the stress depends on the strain in the z direction (it must be zero), and nonzero thermal strain in the z direction will therefore have an effect which needs to be incorporated in the reduced constitutive equation.

The relationship between the normal stresses and the normal strains may be written for the plane strain model as a subset of the full three-dimensional relation

$$\begin{bmatrix} \sigma_x \\ \sigma_y \\ \sigma_z \end{bmatrix} = [D]_{(1:3),(1:3)} \left(\begin{bmatrix} \epsilon_x \\ \epsilon_y \\ \epsilon_z = 0 \end{bmatrix} - \begin{bmatrix} \epsilon_{\Theta x} \\ \epsilon_{\Theta y} \\ \epsilon_{\Theta z} \end{bmatrix} \right). \tag{15.12}$$

Since we are not interested directly in σ_z, and since $\epsilon_z = 0$, we may write

$$\begin{bmatrix} \sigma_x \\ \sigma_y \end{bmatrix} = [D]_{(1:2),(1:2)} \left(\begin{bmatrix} \epsilon_x \\ \epsilon_y \end{bmatrix} - \begin{bmatrix} \epsilon_{\Theta x} \\ \epsilon_{\Theta y} \end{bmatrix} \right) - [D]_{(1:2),3}\epsilon_{\Theta z}. \tag{15.13}$$

Grouping the strains with the same meaning leads to

$$\begin{bmatrix} \sigma_x \\ \sigma_y \end{bmatrix} = [D]_{(1:2),(1:2)} \left(\begin{bmatrix} \epsilon_x \\ \epsilon_y \end{bmatrix} - \begin{bmatrix} \epsilon_{\Theta x} \\ \epsilon_{\Theta y} \end{bmatrix} - [D]_{(1:2),(1:2)}^{-1}[D]_{(1:2),3}\epsilon_{\Theta z} \right), \tag{15.14}$$

where

$$-\begin{bmatrix} \epsilon_{\Theta x} \\ \epsilon_{\Theta y} \end{bmatrix} - [D]_{(1:2),(1:2)}^{-1}[D]_{(1:2),3}\epsilon_{\Theta z},$$

represents an effective in-plane thermal strain.

15.8   Examples

15.8.1   *Thermal strains in a bimetallic assembly*

Consider an assembly of two thin metal slabs. The inset is of aluminum, while the outer plate is of steel: see Fig. 15.9. The assembly is exposed to an increase of 70°C from the stress-free reference state. Of interest are the deformations produced by the unequal coefficients of thermal expansion.

Due to the thinness of the plate, we consider plane stress an adequate approximation. Because of two-way symmetry, only a quarter of the plate is discretized.

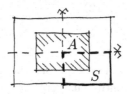

Fig. 15.9   Bimetallic assembly with thermal strain load.

The Matlab script `alusteel`[5] solves the problem with triangular adaptive meshes. The initial phases are omitted for brevity, we just mention that two sets of properties, materials, and finite element blocks need to be created because the assembly consists of two distinct materials.

The calculation proceeds by assuming that the thermal strains are generated by a temperature distribution described by a field. Therefore, such a field is created, and all its degrees of freedom are set to the prescribed temperature.

```
0063 dT = field(struct ('name',['dT'], 'dim', 1, ...
0064     'data',zeros(length(fens),1)+70));
```

Both finite element blocks contribute to the stiffness matrix, and the assembly is performed on the concatenated arrays of element stiffness matrices produced for these two blocks. The thermal strain loads are again computed for each block separately, and assembled into the global load vector.

```
0066 K = start (sparse_sysmat, get(u, 'neqns'));
0067 K = assemble (K, cat(2,stiffness(febs, geom, u),...
          stiffness(feba, geom, u)));
0068 % Load
0069 F = start (sysvec, get(u, 'neqns'));
0070 F = assemble (F, cat(2,
          thermal_strain_loads(feba, geom, u, dT),...
0071      thermal_strain_loads(febs, geom, u, dT)));
```

It might be of interest to point out how to visualize a particular stress component. The first step is to create the graphic viewer, class `graphic_viewer`.

---

[5] Folder: SOFEA/examples/thermo_mechanical

The viewport is reset with the method `reset`.

```
0077 gv=graphic_viewer;
0078 gv=reset (gv,struct ('limits', [0, 100, -8, 40]));
0079 set(gca,'FontSize', 12)
0080 cmap=jet;
0081 cmpn=3;
```

Two fields are created by interpolating the stresses from the integration points to produce continuous representations by estimating stress at the finite element nodes. This is inherently tricky, since we know the stress is in general discontinuous at the nodes. The method `field_from_integration_points` of the class `feblock_defor_ss` uses inverse-distance interpolation. Two distinct fields are used, because the elastic properties of the two materials are different along their common interface.

```
0082 flda = field_from_integration_points(feba, geom, u, dT,
                'Cauchy', cmpn);
0083 flds = field_from_integration_points(febs, geom, u, dT,
                'Cauchy', cmpn);
0084 nvalsa=get(flda,'values');
0085 nvalss=get(flds,'values');
0086 nvalmin =min(min(nvalsa),min(nvalss));
0087 nvalmax =max(max(nvalsa),max(nvalss));
```

An object to map values of the stress to colors, `data_colormap`, is created and initialized with the range of the stress. The stress is then mapped to another field, `colorfield`, whose nodal parameters are colors. The computed color field is used to draw the geometric cells of the block `feba`, using the magnified (scaled) displacement field u so that the deformations are readily distinguishable. The same process is then repeated for the block `febs`.

```
0088 dcm=data_colormap(struct ('range',[nvalmin,nvalmax],
          'colormap',cmap));
0089 colorfield=field(struct ('name', ['colorfield'],
          'data',map_data(dcm, nvalsa)));
0090 draw(feba, gv, struct ('x',geom,'u', scale*u,
          'colorfield',colorfield, 'shrink',1));
0092 colorfield=field(struct ('name', ['colorfield'],
```

```
        'data',map_data(dcm, nvalss)));
0093 draw(febs, gv, struct ('x',geom,'u', scale*u,
        'colorfield',colorfield, 'shrink',1));
```

The problem is solved for four different meshes, each time scaling the element size down by a factor of 2. The shear stress is shown in Fig. 15.10. As can be seen, the stress peaks in the vicinity of the corner in the assembly where the two materials meet. We realize that the reentrant corner in the steel piece is going to generate a stress singularity. To compute the largest stress then becomes futile: it tends to infinity, and any subsequent refinement would only tend to cost more without any genuine improvements. However, in any real design the corner would not really be sharp, there would be a radius to take care precisely of such stress concentrations.

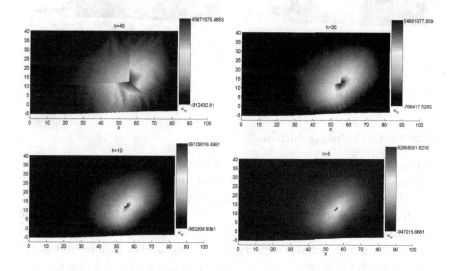

Fig. 15.10　Shear stress in the bimetallic assembly.

The Matlab script alusteelround [6] solves for deformation and stress in a modified geometry, with the reentrant corner rounded to remove the stress singularity. The solution is displayed in Fig. 15.11, and upon closer inspection we can verify that the shear stress is now convergent.

---

[6]Folder: SOFEA/examples/thermo_mechanical

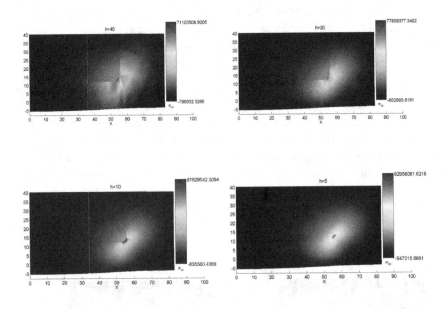

Fig. 15.11   Shear stress in the bimetallic assembly with rounded corner.

15.8.2   *Orthotropic balloon*

In this example we consider an elastic balloon, with orthotropic properties induced by reinforcing fibers along circles in planes perpendicular to the axis of symmetry and centered on the axis of symmetry: see Fig. 15.12. The balloon is loaded by internal pressure. The model is based on the axial symmetry of the expected deformation. The modeled section is hatched, as we use not only the axial symmetry, but also symmetry with respect to the plane through the center of the ball perpendicular to the axis of symmetry.

The mesh is created in a rectangular domain, with the first dimension between zero and the thickness, and the second corresponding to the angle (from zero to the total angle swept out between the axes x and y, which is $\pi/2$). Note that as an option we pass the flag `'axisymm'` to indicate the two-dimensional model reduction type.

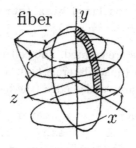

Fig. 15.12 Orthotropic balloon.

```
0021 [fens,gcells]=block2d(rex-rin,pi/2,5,20,
        struct('axisymm',true));
```

In order to apply the pressure boundary condition on the inside of the balloon, we extract the boundary of the block, and we select the corresponding line segment cells.

```
0022 bdry_gcells = mesh_bdry(gcells, struct('axisymm', true,
        'other_dimension', 0.0001));
0023 icl = gcell_select(fens, bdry_gcells,
        struct('box', [0,0,0,pi/2],'inflate',rin/100));
```

The preparation of the mesh is concluded by locating the nodes correctly in the x, y coordinates:

```
0024 for i=1:length (fens)
0025     xy=get (fens(i),'xyz');
0026     r=rin+xy(1); a=xy(2);
0027     xy=[r*cos(a) r*sin(a)];
0028     fens(i)=set(fens(i),'xyz', xy);
0029 end
```

The material model is based on the orthotropic property class, property_linel_ortho, but we choose the material constants so that in the x, y plane the material is isotropic, and the material is therefore effectively transversely isotropic.

```
0031 prop = property_linel_ortho(struct(
        'E1',E1,'E2',E2,'E3',E3,...
0032     'nu12',nu12,'nu13',nu13,'nu23',nu23,...
0033     'G12',G12,'G13',G13,'G23',G23));
```

```
0035 mater=mater_defor_ss_linel_biax (struct('property',prop,
0036     'reduction','axisymm'));
```

Two finite element blocks are created: `feb` collects the finite elements that represent a cross-section (quadrilaterals Q4), and `efeb` for the finite elements L2 along the pressure-loaded edge (the inside of the balloon).

```
0038 feb = febf (struct ('mater',mater, 'gcells',gcells,...
0039     'integration_rule',gauss_rule(2,integration_order)));
0040 efeb = febf (struct ('mater',mater,
         'gcells',bdry_gcells(icl),...
0041     'integration_rule',gauss_rule (1, 2)));
```

The essential boundary conditions are applied in the form of rollers on the axis of symmetry, and on the transverse plane of symmetry.

```
0048 % First the plane of symmetry
0049 ebc_fenids=fenode_select (fens,
         struct('box',[0 rex 0 0],'inflate',rex/10000));
0050 ebc_prescribed=ones(1,length (ebc_fenids));
0051 ebc_comp=ebc_prescribed*0+2;
0052 ebc_val=ebc_fenids*0;
0053 u = set_ebc(u,ebc_fenids,ebc_prescribed,ebc_comp,ebc_val);
0054 % The axis of symmetry
0055 ebc_fenids=fenode_select (fens,
         struct('box',[0 0 0 rex],'inflate',rex/10000));
0056 ebc_prescribed=ones(1,length (ebc_fenids));
0057 ebc_comp=ebc_prescribed*0+1;
0058 ebc_val=ebc_fenids*0;
0059 u = set_ebc(u,ebc_fenids,ebc_prescribed,ebc_comp,ebc_val);
0060 u   = apply_ebc (u);
```

The force intensity object of class `force_intensity` is given a function handle that corresponds to the internal pressure applied radially. The method `distrib_loads` integrates the force intensity over the block `efeb` of the edges on the internal surface, producing an array of element-load vectors. Note the 2 (last argument of `distrib_loads`): it indicates the manifold dimension to be used in the integration of the distributed loads; 2 for a surface.

```
0068 fi=force_intensity(struct('magn',@(x) (p*x'/norm(x))));
0069 F = start (sysvec, get(u, 'neqns'));
```

```
0070 F = assemble (F, distrib_loads(efeb, geom, u, fi, 2));
```

Figure 15.13 displays the deformed shape, which due to the reinforcing fibers running in planes perpendicular to the axis of symmetry assumes the general resemblance of a football. The circumferential stress is displayed on the deformed shape, while the undeformed shape is shown as shrunk outlines of the cross-section elements.

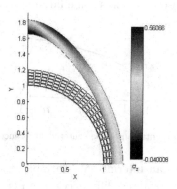

Fig. 15.13   Orthotropic balloon.

15.9   Transient dynamic analysis

In this section we introduce the explicit variant of the Newmark algorithm, the **centered difference** integrator. It will be convenient to derive the algorithm as if we were thinking of the motion of a particle since to make it work for the finite element model, we just replace scalars with matrices.

15.9.1   *Centered difference time stepping*

The starting point is the well-known formula for the approximation of the second order derivative of the displacement

$$\ddot{u} \approx \frac{u_{n+1} - 2u_n + u_{n-1}}{\Delta t^2} = a_n , \qquad (15.15)$$

where we introduce a label for the approximation, a_n, which is the so-called *algorithmic acceleration*; also, we use the time step $\Delta t = t_n - t_{n-1}$. Another way of computing this acceleration is from the equation of motion

(Newton's equation)

$$a_n = M^{-1}f(t_n, u_n) = M^{-1}f_n \,, \qquad (15.16)$$

where f_n is the resultant force. Since we assume u_n and u_{n-1} as the displacements of the particle at time instants t_n and t_{n-1} are known, we can use these two equations to solve for the next displacement, u_{n+1}. However, it is slightly inconvenient to have to keep track of the previous displacement, u_{n-1}, especially because it requires some fiddling in the first step.

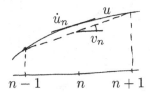

Fig. 15.14   Graphical representation of the centered difference definition of the velocity.

A remedy is to use an intermediate variable, the *algorithmic velocity*. We define, again using centered differences (see Fig. 15.14),

$$v_n = \frac{u_{n+1} - u_{n-1}}{2\Delta t} \,. \qquad (15.17)$$

Therefore, we can express

$$u_{n-1} = -2\Delta t v_n + u_{n+1} \,, \qquad (15.18)$$

and substituting into (15.15), write the displacement

$$u_{n+1} = u_n + \Delta t v_n + \frac{\Delta t^2}{2} a_n \,. \qquad (15.19)$$

It remains to determine how to push forward the algorithmic velocity. Combining the two Eqs. (15.18) and (15.19), written for the time instants t_{n+2}, t_{n+1} and t_n

$$v_{n+1} = \frac{u_{n+2} - u_n}{2\Delta t} \,, \quad u_{n+2} = u_{n+1} + \Delta t v_{n+1} + \frac{\Delta t^2}{2} a_{n+1} \,, \qquad (15.20)$$

allows us to express u_{n+2} in two ways, and obtain therefore an equation for v_{n+1}

$$v_{n+1} = v_n + \frac{\Delta t}{2}(a_n + a_{n+1}) \,. \qquad (15.21)$$

To summarize, the **centered difference algorithm** now reads

$$\text{Given:} \quad u_n, v_n, a_n = M^{-1}f(t_n, u_n) \tag{15.22}$$

Compute:

$$u_{n+1} = u_n + \Delta t v_n + \frac{\Delta t^2}{2}a_n$$

$$a_{n+1} = M^{-1}f(t_{n+1}, u_{n+1})$$

$$v_{n+1} = v_n + \frac{\Delta t}{2}(a_n + a_{n+1})$$

As shown for instance in Reference [Hughes (2000)], this algorithm is only conditionally stable, meaning that it blows up for time steps longer than a **critical time step**

$$\Delta t_{\text{crit}} = \frac{2}{\max \omega_h} = \frac{\min T_h}{\pi}, \tag{15.23}$$

where $\max \omega_h$ is the highest frequency of vibration in the discrete system, and $\min T_h$ is the associated period. Cruder estimates of Δt_{crit} are available also from the vibration frequencies of the individual elements in the mesh, or as the smallest time interval needed to travel the distance between finite element nodes by an elastic wave.

15.9.2   *Example: stress wave propagation*

In this section, we will use the centered difference algorithm to obtain a transient solution for the problem of a wave propagating through a cylindrical rod in which there is a spherical cavity. We would expect to see reflections of the stress wave off the free surface of the cavity. Refer to Fig. 15.15. The Matlab script `pressrecho`[7] solves the problem as axially

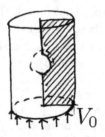

Fig. 15.15   Cylindrical rod with the spherical cavity.

---

[7]Folder: SOFEA/stress/2-D

symmetric, using linear triangles. The issue of convergence is not being addressed, but in any serious investigation it certainly should be: as we are well aware by now, the linear triangles are not very accurate.

The mesh is created in the axially symmetric domain with the automatic mesh generator. Note that an options structure is being passed to make all the generated triangles axisymmetric: `struct('axisymm', true)`.

```
0014 [fens,gcells,groups,edge_gcells,edge_groups]=
             targe2_mesher({...
0015    ...
0025    }, 1.0,struct('axisymm', true));
```

The property, material, and finite element block, geometry field and displacement field objects are created in the usual way. Displacement boundary condition is applied only along the axis of symmetry.

Next, the initial velocity v field is created by copying the displacement field; since it is a *copy* of the displacement field, the numbers of equations are now available in the velocity field. These are used later to retrieve the system vector initialized in the proper places with the initial velocity. In this example, the initial velocity is imparted to the nodes on the bottom cross-section to simulate for instance an impact.

```
0050 v = u;% copy displacement field to velocity
0051 fenids=fenode_select (fens,struct('box',[0 15 -25 -25],
             'inflate',1/100));
0052 for j=1:length(fenids)
0053     xy= get (fens(fenids(j)),'xyz');
0054     v = scatter(v,fenids(j),[0, 1]*vmag);
0055 end
```

The stiffness and mass matrices are assembled: note that the mass is assembled as a diagonal matrix. That affords the best efficiency, since all the solutions of the type `A0 =Mmat\(-Kmat*U0)` are then very quick indeed. The diagonalization is effected by lumping the element mass matrices using the row-sum technique (line 0063).

```
0056 % Assemble the system matrix
0057 K = start (sparse_sysmat, get(u, 'neqns'));
0058 K = assemble (K, stiffness(feb, geom, u));
0059 M = start (sparse_sysmat, get(u, 'neqns'));
0060 ems=mass(feb, geom, u);
```

```
0061 for j=1:length(ems)
0062     Me=get(ems(j),'mat');
0063     ems(j)=set(ems(j),'mat',diag(sum(Me)));
0064 end
0065 M = assemble (M, ems);
```

The initial values for the integration are established next.

```
0068 Kmat = get(K,'mat');
0069 Mmat = get(M,'mat');
0070 U0 = gather_sysvec(u);
0071 V0= gather_sysvec(v);
0072 A0 =Mmat\(-Kmat*U0);
```

The critical time step is calculated by computing the largest natural frequency of the discrete system. This might not be a good idea for very large meshes, since the eigenvalue solution takes time.

```
0074 o2=eigs(Kmat,Mmat,1,'LM');
0075 dt= 0.899* 2/sqrt(o2)
```

The loop really transcribes the algorithm (15.22) more or less literally. A minor addition is the re-initialization after each step (lines 0088-0090).

```
0083 while t <tfinal
0084     t=t+dt;
0085     U1 = U0 +dt*V0+(dt^2)/2*A0;
0086     A1 = Mmat\(-Kmat*U1);
0087     V1 = V0 +dt/2* (A0+A1);
0088     U0 = U1;
0089     V0 = V1;
0090     A0 = A1;
0091     ...
0106     end
0107 end
```

Figure 15.16 illustrates the pressure at equally spaced time intervals. Given the fairly limited resolution (approximately 30 elements radially), the results are far from smooth, but still afford a qualitative picture of a pressure echo.

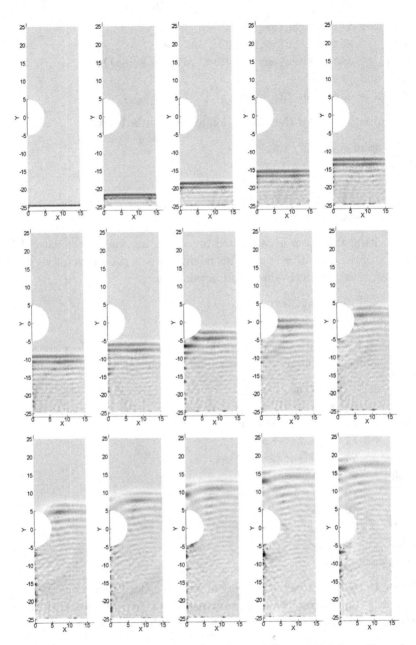

Fig. 15.16   Propagation of a stress wave through an axially symmetric bar with a spherical cavity. Displayed is the pressure at equally spaced time intervals.

Exercises

(1) How would damping forces be included in the centered difference time stepping? Consider the equation of motion $M\ddot{u} + C\dot{u} + Ku = L$, where the damping forces are velocity-proportional (viscous).

Chapter 16

Consistency + Stability = Convergence

Let us assume the elasticity problem we wish to solve has a nice and smooth "exact" solution. What are the conditions for a finite element model to converge to this exact solution in the limit of the mesh size approaching zero? Even though we are never going to be able to reach this limit except in special situations, we need to know that it is possible to approach it.

For the simple finite element models that are considered in this book, there are two sufficient conditions for convergence to occur: *consistency* and *stability*.

16.1 Consistency

Consistency is really just a front for two individual requirements: Completeness + Compatibility. For the sake of the argument the following discussion will be centered around the model of elasticity (statics), but the conclusions may be extended to other problems more or less readily.

16.1.1 *Completeness*

A key quantity in the elasticity model are the strains. The completeness requirement aims to ensure that all possible constant strains within an element are represented exactly. (When we say all possible constant strains, we also mean zero strains – rigid body motion.) Let us build on our intuition: imagine being able to observe the strains on a deforming structure as a continuous field (Fig. 16.1). When we observe the whole structure, the level surfaces of the strains are very complex, but as we look closer and closer at the structure, the variations in the strains become less and less distinguishable. (This is the same principle as in the truncated Taylor

Fig. 16.1   Zooming in on the strain with a microscope.

series: the smaller the excursion from a given point, the more important the lower-order terms will become. Looking at the point itself, only the constant term remains.) In other words, looking close enough we see material experiencing constant strain. *If this material was enclosed in a finite element, we would want the finite element to be able to represent this state.*

Since the strains are obtained as combinations of the first derivatives of the displacements, we conclude that for the element to be able to represent constant strain, it must be able to reproduce *linear polynomials exactly.* Clearly, the simplex elements (L2, T3, T4) satisfy this requirement by construction. However, it is easy to verify that *all* the iso-parametric elements discussed in this book reproduce exactly linear functions: Assume a linear function in this form

$$f(\boldsymbol{x}) = \boldsymbol{a}\boldsymbol{x} + b \, ,$$

where $\boldsymbol{a}$ is a constant row matrix, and b a constant. Now write a finite element interpolation of the function f as

$$\Pi_h f(\boldsymbol{x}) = \sum_{k=1}^{M} N_k(\boldsymbol{x}) f_k \, .$$

Provided the degrees of freedom are set as

$$f_k = \boldsymbol{a}\boldsymbol{x}_k + b \, ,$$

where $\boldsymbol{x}_k$ are the position vectors of the nodes, we have

$$\Pi_h f(\boldsymbol{x}) = \sum_{k=1}^{M} N_k(\boldsymbol{x}) f_k = \sum_{k=1}^{M} N_k(\boldsymbol{x}) \left[\boldsymbol{a}\boldsymbol{x}_k + b \right] \, .$$

But due to the partition of unity property (6.15) and to the interpolation of the Cartesian coordinates (6.16) we have

$$\Pi_h f(\boldsymbol{x}) = \sum_{k=1}^{M} N_k(\boldsymbol{x}) \left[a\boldsymbol{x}_k + b \right] =$$

$$a \sum_{k=1}^{M} N_k(\boldsymbol{x})\boldsymbol{x}_k + b \sum_{k=1}^{M} N_k(\boldsymbol{x}) = a\boldsymbol{x} + b = f(\boldsymbol{x}) .$$

Therefore, all the finite elements that satisfy (6.15) and (6.16) are complete in the sense discussed in this paragraph.

16.1.2 *Compatibility*

The elasticity mathematical model describes a material which does not separate or fracture. Therefore, also the finite element model must preserve the continuity of the material. The basis functions within an element describe how particles within the confines of the element move. Since all the basis functions we discussed were continuous inside the element, the compatibility requirement is satisfied inside the element. However, we also need to insure that the material does not experience overlaps or fissures where the individual elements meet. Therefore, we formulate the compatibility requirement along the edges (faces or vertices): all elements meeting along an edge (along a face, or at a vertex) of the mesh must describe the displacements in the same way. Thus, mixing together three-node and six-node triangles could result in gaps opening in the material (compare with Fig. 16.2), because along an edge the displacement is a linear function from one side, and a quadratic function from the other. And if such gaps proliferated in the mesh, there would be a portion of energy improperly accounted for, since these gaps do not exist in the mathematical model of the structure.

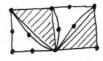

Fig. 16.2 Incompatible arrangement of elements.

16.2 Stability

Stability determines the uniqueness of the finite element solution. The mathematical model of elasticity does not admit solutions for the displacement that do not produce positive energy other than rigid body motion. (Provided the material is stable: the material stiffness matrix must be positive definite.) Therefore, the finite element model must not have built-in **zero-energy modes** other than rigid-body modes. However, in order for most linear algebra algorithms and solvers to work, we would remove all rigid body modes before a solution is attempted on the level of the finite element system: the global stiffness matrix should not be singular.

Therefore, the stability requirement is essentially reduced to demanding that the *element stiffness matrices* be of *proper rank*, which means of rank that allows only for zero-energy modes in the form of rigid body motion. The element stiffness matrix K_e which links together n degrees of freedom (all the displacement components at all the nodes) should have the rank

$$\mathrm{rank} K_e = n - n_{\mathrm{RBM}} \ ,$$

where n_{RBM} is the number of rigid body modes the element possesses (=6 in 3-D, and so on). To dig deeper, we recall Eq. (12.44), which we supplement with numerical quadrature

$$K_e = \int_{V_e} B^{eT} D B^e \ \mathrm{d}V \approx \sum_k B^e(\xi_k)^T D(\xi_k) B^e(\xi_k) J(\xi_k) w_k \ . \quad (16.1)$$

Assuming the full rank of the material stiffness matrix (positive definite!), the requirement that the element stiffness matrix have a proper rank may be translated into a counting procedure for the integration points: each quadrature point adds the matrix

$$B^e(\xi_k)^T D(\xi_k) B^e(\xi_k) J(\xi_k) w_k \ ,$$

which has the rank of $\mathrm{rank} D(\xi_k)$, with the provisions $J(\xi_k) \neq 0$, and $w_k \neq 0$. If each quadrature point increases the rank of K_e by $\mathrm{rank} D(\xi_k)$, the requisite number of integration points will be

$$n_{qp} \geq \frac{n - n_{\mathrm{RBM}}}{\mathrm{rank} D} \ . \quad (16.2)$$

For instance, for the quadratic tetrahedron T10, the number of quadrature points should be

$$n_{qp} \geq \frac{30 - 6}{6} = 4 \,,$$

and that is indeed satisfied by the four-point rule from Table 9.2.

It should be noted that the Jacobian could cause some trouble. Figure 16.3 shows a few possible shapes of a quadratic triangle obtained by shifting the mid-side nodes. The center triangle represents the situation in which we wish to adjust the location of a node in order to better approximate curved boundaries. That is typically done, and moderate excursions of the nodes typically do not lead to major difficulties. The triangle on the left is produced by shifting the mid-side nodes inwards excessively so that the surface of the triangle is turned inside out, resulting in a negative Jacobian in the corner regions (zero Jacobian where the edges intersect). The triangle on the right is obtained by shifting the mid-side node along the edge towards the corner. Beyond a certain point, the Jacobian again becomes negative around the corner. (To realize what is happening, note that a parabolic arc needs to be passed through the three nodes along an edge.) Negative or zero Jacobians are a problem, both in terms of the rank of the element stiffness matrix, and in terms of the interpretation of the strains produced on such elements.

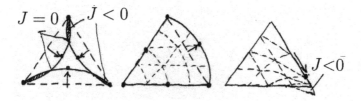

Fig. 16.3 Shapes of quadratic triangles produced by moving mid-side nodes away from the average of the locations of the corner nodes.

16.2.1 *Conclusion*

The following formal equality has been coined for finite difference methods, but will also work for the finite elements here:

Consistency + Stability = Convergence.

We have shown, that convergence for properly defined quantities (not for infinite stresses!) is available with the methods discussed in this book: they possess the property of consistency, and with proper integration, they may be endowed with stability.

Some of the above requirements may be mollified or relaxed, and there may be good reasons for doing so: for instance, improving performance on coarse meshes by relaxing the constraints on the compatibility. However, as a general guide towards convergent finite element approximations, the above dictum will be a safe starting point.

Exercises

(1) Consider a T6 element which is mapped to the physical space with the identity map, $x = \xi$, $y = \eta$. Move the node 6 to $x = \overline{\xi}$, $y = 1/2$. Find $\overline{\xi}$ such that the Jacobian becomes zero in at least one point.

(2) Deduce how many Gauss integration points will be needed to ensure proper rank of the conductivity matrix for the element Q4.

(3) Compute the rank of the conductivity matrix of a well-shaped T6 element using a one-point quadrature. Is the number of (nearly) zero eigenvalues consistent with the argument in Eq. (16.2)?

(4) Deduce how many Gauss integration points will be needed to ensure proper rank of the conductivity matrix for the element Q8.

Bibliography

Barber, J. R. (1999). Elasticity, Kluwer Academic Publishers, reprint edition.

Blevins, R. D. (2001). Formulas for Natural Frequency and Mode Shape, Krieger publishing company, reprint edition.

Bogacki, P. and L. F. Shampine (1989). A 3(2) pair of Runge-Kutta formulas, Appl. Math. Letters, Vol. 2, pp 1-9.

Cameron, A. D. and J.A. Casey, G.B. Simpson (1994). Benchmark Tests for Thermal Analyses, NAFEMS Documentation.

Graff, K. F. (1991). Wave motion in elastic solids. Dover publications, New York.

Hughes, Thomas J. R. (2000). The finite element method. Linear static and dynamic finite element analysis. Dover publications, New York.

Hyer, M. W. (1998). Stress Analysis of Fiber-Reinforced Composite Materials, WCB/McGraw-Hill.

Infolytica (2005). http://www.infolytica.com/en/coolstuff/ex0047/

Lienhard IV, John H. and John H. Lienhard V (2005). A Heat Transfer Textbook, 3rd edition, http://web.mit.edu/lienhard/www/ahtt.html.

Matlab online documentation, Release 14.
http://www.mathworks.com/access/helpdesk/help/helpdesk.html

Moler, C. (2004). Numerical Computing with MATLAB, Electronic edition: The MathWorks, Inc., Natick, MA, 2004. http://www.mathworks.com/moler Print edition: SIAM, Philadelphia, 2004.
http://ec-securehost.com/SIAM/ot87.html

Pilkey, W. D. (1997). Peterson's Stress Concentration Factors. John Wiley & Sons, second edition.

Sokolnikoff, I. S. (1983). Mathematical Theory of Elasticity. Krieger publishing company. Reprint edition 1983.

Timoshenko, S. P. (1983). History of Strength of Materials. Dover, 1983.

Timoshenko, S. P. and S. Woinowsky-Krieger (1959). Theory of Plates and Shells. McGraw-Hill, 1959.

Zienkiewicz, O. C. and R. L. Taylor (1989). The finite element method. Vol. I. Basic formulations and linear problems. London: McGraw-Hill.

Index